Organic Chemistry

How Organic Chemistry Works

Studymates

Algebra
Better English
Better French
Better French 2
Better German
Better Spanish
Calculus
Chemistry A/s chemistry Explained
Chemistry: Chemistry Calculations Explained
Genetics
Hitler and Nazi Germany
Lenin, Stalin and Communist Russia
Perfect Grammar
Plant Physiology
Poetry
Religion
Shakespeare
Social Anthropology
Starting Statistics
Study skills
The Academic Essay
The Theatre Makers
The War Poets
Understanding Film and Literature
Understanding Maths
Warfare

Many other titles in preparation

Studymates

Helping You to Achieve

Organic Chemistry

How Organic Chemistry Works

Dr Aleyamma Ninan

Second edition
Published by Studymates 2009, reprinted 2011, 2013

ISBN 978 1 84285 090 9

Contents

Preface

This book aims to provide a concise introduction to organic chemistry at university/higher education level whether you are doing it as a main subject, or as part of a biology, medical or environmental science degree. It should also help you if you haven't done A-level chemistry, since it covers the basic principles of organic chemistry. This book has been deliberately written in a clear, brief, lecture-note style. There are many other textbooks aimed at university/degree level, often vast in scope but sometimes rather unwieldy.

The basic skills that you need to develop as a student of organic chemistry include: learning how to write the structure of organic compounds, the IUPAC (International Union of Pure and Applied Chemistry) system of naming organic compounds, the mechanisms of organic reactions, and organic compound stereochemistry. These important topics are treated in a simple and easy to understand manner. I have also introduced topics like organic synthesis and spectroscopic analysis, which are important tools for modern organic chemists.

This book is intended to be used as a study and revision guide, covering all the essentials of the subject. You can use it alongside a standard textbook which should give you the further details.

I would like to thank David Brown and Golda Ninan for reading the entire text and giving suggestions, Harry Hartel and Chris Cryer for providing me with spectral data, and Silva Ninan and Matthew Philip for helping with computer skills. Finally, I would like to thank my family for their support and encouragement throughout the period of writing this book.

A Ninan
aninan@studymates.co.uk

1

Structure and Bonding

One-minute summary – The main type of force of attraction or bonding between atoms in simple organic compounds is the covalent bond. The formation of a covalent bond can be described in terms of the sharing of two unpaired electrons between two atoms or in terms of the overlapping of two atomic orbitals. The atomic orbitals that take part in the bond formation can be pure atomic orbitals or hybrid atomic orbitals. The bonds present in most organic compounds are not purely covalent, but have some degree of polar or ionic character because the shared electron pair in the bond is attracted and kept closer to one nucleus than to the other. This is because of the difference in electronegativity between the two atoms. The aim of this chapter is to present the modern approach to covalent bond formation. The following terms, necessary for the understanding of covalent compounds and their nature, are explained below.

- ▶ atomic orbitals, hybrid atomic orbitals and molecular orbitals
- ▶ formation of a covalent bond by the overlapping of orbitals
- ▶ sigma and pi covalent bonds
- ▶ shape and symmetry of simple molecules
- ▶ bond length and bond angle
- ▶ electronegativity of elements and polarity of bonds.

1.1 What are atomic orbitals?

An **atom** is made up of protons, electrons and neutrons. The positively charged protons and the neutral neutrons are packed together to form the nucleus of the atom. The negatively charged electrons are distributed around the nucleus in atomic orbitals of various energy levels or shells. The energy levels are called K, L, M – and numbered 1, 2, 3 – starting from the innermost one.

An **atomic orbital** is the region of space where an electron is found. Picture each electron in an atom moving, vibrating and spinning constantly around the nucleus forming a negatively charged electron cloud. The shape of the atomic orbital is the shape of the electron cloud. The atomic orbital represents the region where the probability of finding the electron is highest.

There are different types of atomic orbitals designated by the letters *s*, *p*, *d* and *f*. In the first energy level, there is only one orbital, and it is an *s* orbital. In the second energy level there are two types of orbitals; *s* and *p* orbitals, in the third energy level, three types; *s*, *p* and *d* orbitals and so on. There is only one *s* orbital in any one energy level, while there are three *p* orbitals and five *d* orbitals. The *s* orbital of the first energy level is called 1*s*, that of the second energy level, 2*s*, the *p* orbitals of the third energy level, 3*p* and so on.

Note that each electron in an atom possesses a certain amount of energy. An electron on a level closer to the nucleus has less energy than an electron on a higher level. An electron needs to gain a certain amount of energy before it can move to a higher level. Similarly, an electron loses energy and falls into a lower level if there is space available.

Energy level	Number of types of orbitals	Name of orbitals	Number of electrons
1 (K-shell)	one	1s	2
2 (L-shell)	two	2s	2
		2p	6
3 (M-shell)	three	3s	2
		3p	6
		3d	10

Table 1.1. Number and type of orbitals.

An *s* orbital is spherical in shape and can be represented by a sphere as shown in Figure 1.1. The nucleus is at the centre of the sphere. The 2*s* orbital is larger than the 1*s* orbital.

The *p* orbital (Figure 1.2) is dumb-bell shaped and consists of two lobes, with the nucleus located between the two. There are three *p* orbitals of equal energy designated p_x, p_y and p_z. They are directed along x, y and z axes and they are perpendicular to one another.

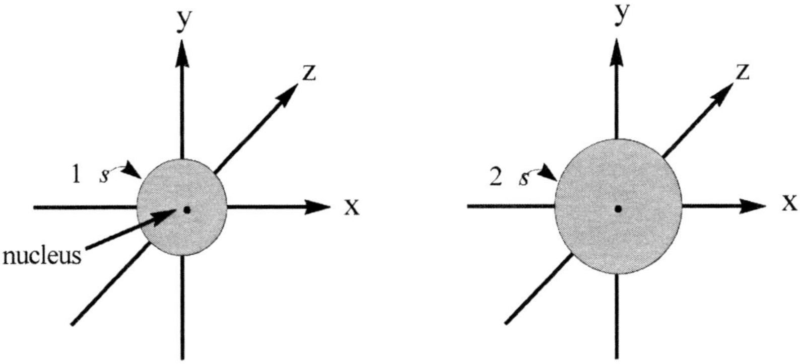

Fig. 1.1. The shape of 1*s* and 2*s* orbitals.

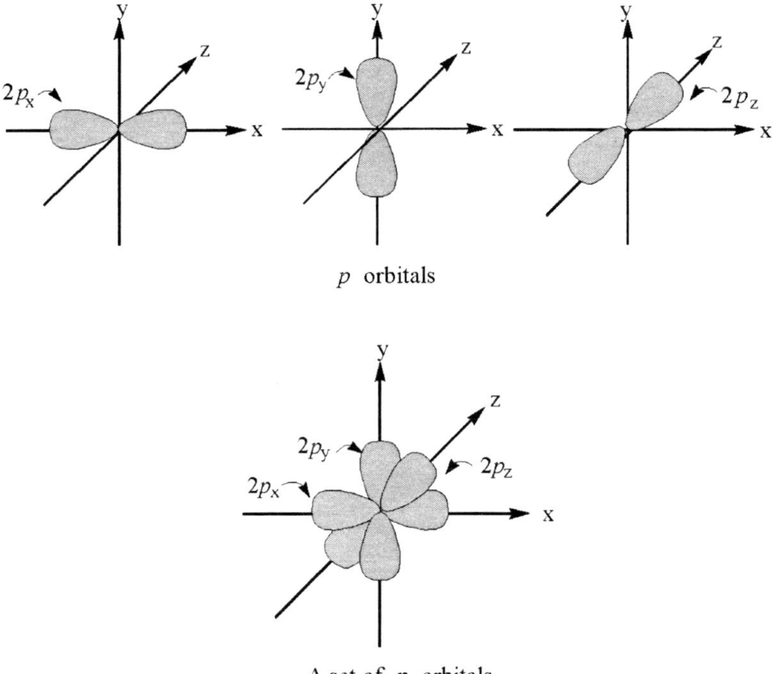

Fig. 1.2. *p* orbitals.

1.2 Writing electronic configuration

What is electronic configuration? How do we write the electronic configurations of elements?

Electronic configuration is the arrangement of available electrons of an atom, in named atomic orbitals. The following rules are helpful for writing the electronic configurations of elements:

1. In an atom, electrons occupy the innermost energy level, filling it before going to the next higher energy level (The *Aufbau* or *building up principle*).

2. An orbital is occupied by a maximum of two electrons (*Pauli exclusion principle*) and these two electrons spin in opposite directions. Thus an *s* orbital is filled by two electrons, the three *p* orbitals of an energy level by six electrons, the five *d* orbitals of an energy level by a maximum of 10 electrons, and so on (Table 1.1).

3. When there is more than one orbital of equal energy value (for example, the three *p* orbitals in an energy level), electrons occupy them singly, before pairing takes place (*Hund's rule of maximum multiplicity*).

Now we can write the electronic configuration of some elements. A hydrogen atom (atomic number, Z = 1) has one electron and it occupies the 1*s* orbital. Its electronic configuration is written as:

H: $1s^1$
(Note that the superscript 1 stands for the number of electrons in the orbital)

A helium atom (Z = 2) has two electrons, both of which remain in the 1*s* orbital. A lithium atom (Z = 3) has three electrons, two of which occupy the 1*s* orbital and the third one, the 2*s* orbital. The electronic configurations of helium and lithium are:

He: $1s^2$
Li: $1s^2\ 2s^1$

Following the above principles, the electronic configurations of carbon

(Z = 6), nitrogen (Z = 7), oxygen (Z = 8) and sodium (Z = 11) are written as:

C: $1s^2\ 2s^2 2p_x^1 2p_y^1$
N: $1s^2\ 2s^2\ 2p_x^1\ 2p_y^1\ 2p_z^1$
O: $1s^2\ 2s^2\ 2p_x^2\ 2p_y^1\ 2p_z^1$
Na: $1s^2\ 2s^2\ 2p^6\ 3s^1$

Notes

1. **Atomic number** is the number of protons or the number of electrons in an atom and can be obtained from the periodic table.
2. The two *p* electrons of carbon are distributed in two of the three available *p* orbitals and the three *p* electrons in nitrogen, in the three *p* orbitals, since all the three *p* orbitals have the same energy.
3. When all the three *p* orbitals of an energy level have been filled, the arrangement can be written as p^6 instead of $p_x^2 p_y^2 p_z^2$.

1.3 Formation of a covalent bond

A **covalent bond** between two atoms in a molecule can simply be described as the attraction between the atomic nuclei and a pair of electrons shared by the two atoms. For example, a hydrogen molecule (H_2) is formed when the two electrons from two hydrogen atoms form **a shared pair**, thus holding the two hydrogen nuclei together.

In a hydrogen chloride (HCl) molecule, an electron from a hydrogen atom and the unpaired electron in the outermost orbit of a chlorine atom (Cl: $1s^2 2s^2 2p^6 3s^2 3p_x^2 3p_y^2 \mathbf{3p_z^1}$) form a shared pair, thus forming a covalent bond.

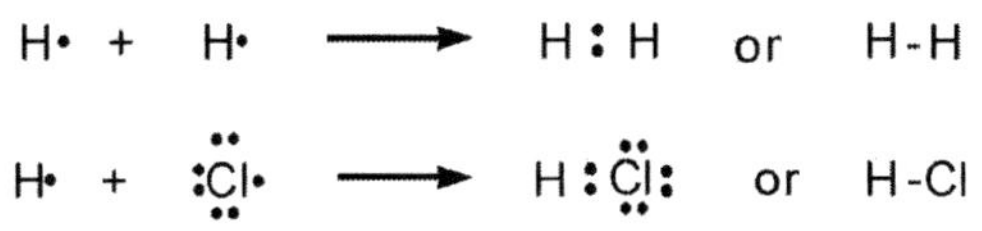

Notes

1. The dots around an atom represent the electrons of the outermost shell of the atom.
2. A dash (-) between two atoms stands for an electron pair or a covalent bond.

Now, let us see how covalent bond formation is explained in terms of the **overlapping of orbitals**. For example, when two hydrogen atoms come close enough, the 1*s* orbital of one atom overlaps or merges with a similar orbital of the other atom to form a molecular orbital. A **molecular orbital** is the region of space in a molecule where electrons are found. A molecular orbital like an atomic orbital is occupied by a maximum of two electrons of opposite spin. The two electrons in the hydrogen molecule spend most of the time in the space between the two nuclei. The shape of the molecular orbital of hydrogen can be represented as:

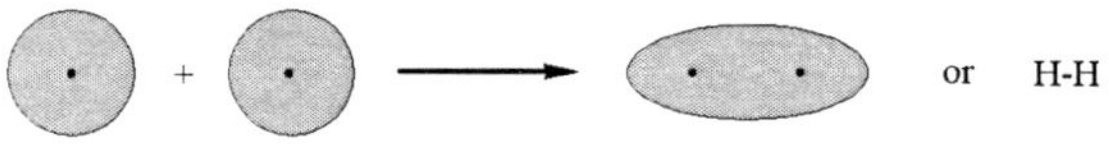

Figure 1.3. The formation of a H_2 molecule.

For hydrogen, the extent of overlapping between two *s* orbitals is high, which means that the bond formed is strong, and such a bond is called a **sigma (σ) covalent bond**. The molecular orbital formed is symmetrical around a line joining the two nuclei.

A hydrogen molecule is more stable than the two individual hydrogen atoms. This is because energy is released during the formation of a hydrogen molecule or, conversely, energy is required to break the bond in a hydrogen molecule. The energy required to break the bond in one mole of diatomic molecules is called the **bond dissociation energy**. The bond dissociation energy of hydrogen is 435.9 kJ mol^{-1}.

Bond length is the distance between the two atomic nuclei which share the bond. The H-H bond length of a hydrogen molecule is 0.074 nm (1 nm = 10^{-9} m).

A hydrogen chloride molecule is formed when the *s* orbital of the hydrogen atom overlaps with the half-filled *p* orbital of the chlorine atom (Figure 1.4). The bond formed by the overlapping of an *s* orbital and a *p* orbital is also a sigma covalent bond since maximum electron density lies about the axis joining the two nuclei. The H-Cl bond dissociation energy is 432.0 kJ mol^{-1} and the H-Cl bond length is 0.127 nm.

The bonds in simple covalent molecules like H-F, H-Br, H_2O, F- and I-I are sigma covalent bonds.

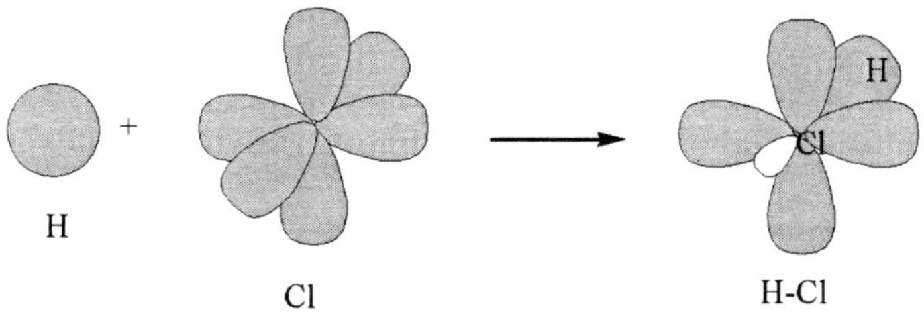

Figure 1.4. The formation of a H-Cl molecule.

1.4 Hybridisation of atomic orbitals

sp^3 hybridisation

Methane, CH_4, is the simplest stable organic molecule and it is a covalent compound. The equivalence of the covalent bonds in methane can be explained in terms of **hybridisation of atomic orbitals**. A carbon atom in its 'ground' state has only two unpaired electrons (C: $1s^2 2s^2$ $\mathbf{2p_x^1 2p_y^1}$) which can form only two covalent bonds. During compound formation one of the two 2*s* electrons of the carbon atom is promoted to the vacant 2*p* orbital. The energy required for exciting or promoting the electron (about 400 kJ mol^{-1}) is available during the bond formation.

The carbon atom in the 'excited' state has four unpaired electrons (C: $1s^2\mathbf{2s^1 2p_x^1 2p_y^1 2p_z^1}$) and can now form four covalent bonds. But since the four unpaired electrons are in different types of orbitals, we expect the bonds in methane to be different. But all the four bonds in methane are equivalent and all the bond angles equal. This is because the *2s* orbital and the three *2p* orbitals undergo hybridisation to form four new orbitals. This is called **sp^3 hybridisation** which means each new orbital has 25% *s* and 75% *p* character. The orbitals formed are called **sp^3 hybrid orbitals**. Each *sp*^{3} hybrid orbital consists of a larger lobe and a smaller lobe pointing in opposite directions. These orbitals are equivalent and the axes of the orbitals are directed towards the corners of a regular tetrahedron.

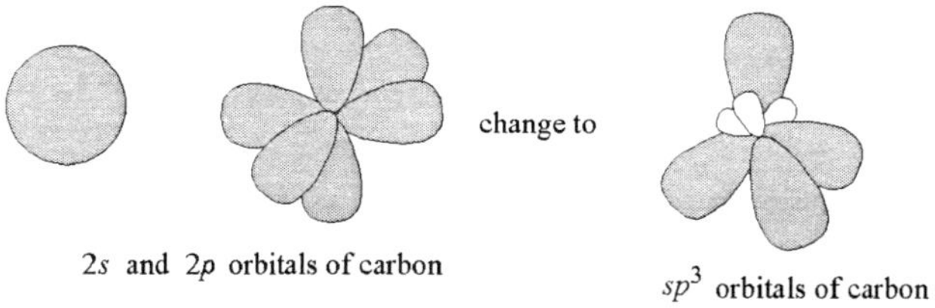

Figure 1.5. *sp*^{3} hybrid orbitals.

In methane, the carbon atom is sp^3 hybridised and the four hybrid orbitals each containing one electron, overlap with four *s* orbitals of four hydrogen atoms to form four sigma covalent bonds. The shape of the methane molecule is tetrahedral, with the carbon atom at the centre and the four hydrogen atoms at the corners of a regular tetrahedron.

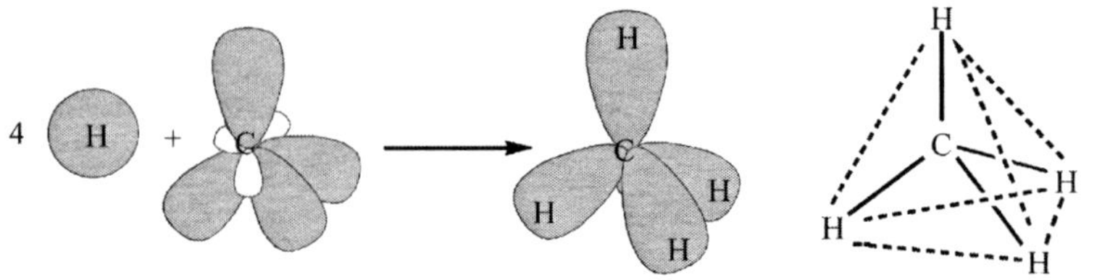

dotted lines show the regular tetrahedron

Figure 1.6. The formation of a methane molecule.

As mentioned earlier, energy is required for exciting an electron from 2*s* to 2*p* orbital. The rise in energy of the system in the excited state is far outweighed by the lowering of energy (stabilization) of the system by the formation of four strong bonds (rather than two weak ones). The average bond dissociation energy of a C-H bond in methane is 412 kJ mol^{-1} and C-H bond length is 0.11 nm.

The angle between two covalent bonds formed from one atom is called the **bond angle**. All the bond angles in methane are equal to 109^0 28' which is the expected value for a tetrahedral arrangement of atoms.

Generally, the carbon atoms in saturated organic compounds are sp^3 hybridised. A **saturated** organic compound is one in which each carbon atom is bonded to four other atoms by covalent bonds. Thus, the carbon atoms in ethane, C_2H_6, ethanol, C_2H_5OH, etc are sp^3 hybridised.

sp^2 hybridisation

Ethene, C_2H_4, is another simple organic compound. There are two covalent bonds between the two carbon atoms in an ethene molecule.

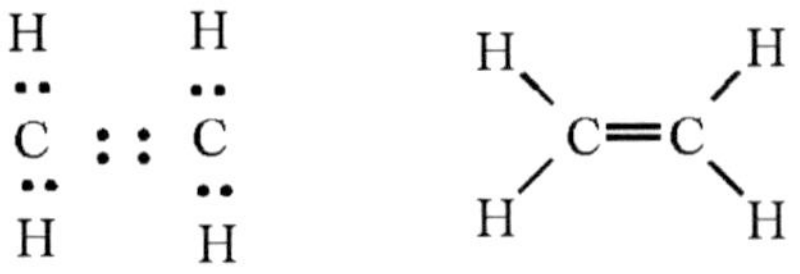

Ethene

A compound which has two or three covalent bonds between two carbon atoms is called an **unsaturated** compound.

The carbon atoms in ethene, C_2H_4, are said to be *sp*^{2} hybridised. The 2*s* orbital and two of the three 2*p* orbitals of the carbon atom (in the excited state) take part in the hybridisation to form three equivalent orbitals called **sp^2 orbitals**. The hybridisation of one *s* orbital and two *p* orbitals is called **sp^2 hybridisation**. One of the three *p* orbitals does not take part in the hybridisation.

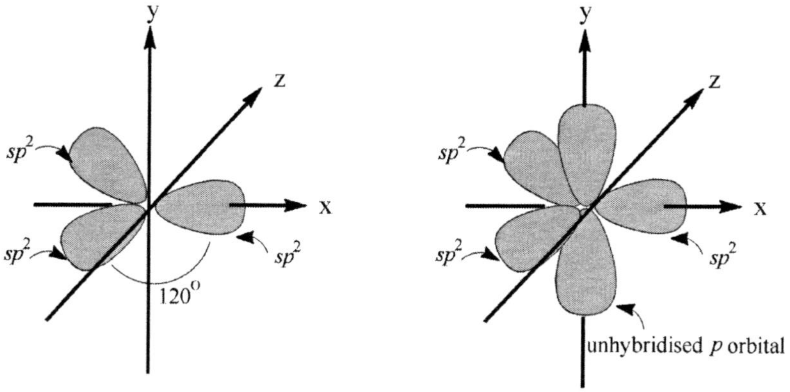

Figure 1.7. *sp*^{2} hybrid orbitals of a carbon atom.

The three *sp*^{2} orbitals are in the same plane and their axes are at an angle of 120^0 to each other. The axis of the unhybridised *p* orbital is perpendicular to the plane of the hybrid orbitals.

In ethene, two of the *sp*^{2} orbitals of each carbon atom overlap with two *s* orbitals of two hydrogen atoms and the third *sp*^{2} orbital of each carbon atom overlaps endwise. All the five bonds formed are σ- bonds and they are in one plane. The two unhybridised *p* orbitals of the two carbon atoms, which are parallel to one another, overlap sidewise to form a weak covalent bond. The extent of sidewise overlapping is less than that of endwise overlapping and such a covalent bond is called a **pi (π) bond**. Thus, there are two types of covalent bonds between the two carbon atoms in ethene, a sigma and a pi bond. A molecule of ethene is flat since the geometry is determined by the *sp*^{2} orbitals of the carbon atoms. The H-C-H and H-C-C bond angles are close to the expected value of 120°. The H-C-H bond angle in ethene is 116.6° and the H-C-C bond angle is 121.7°.

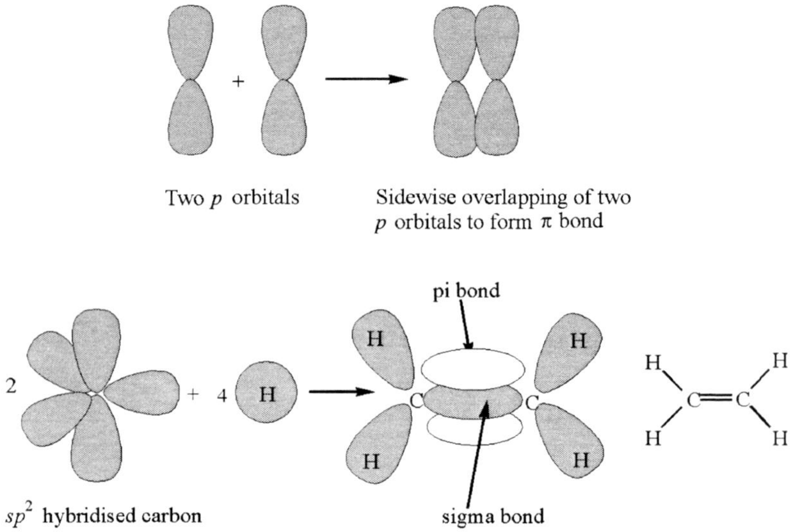

Figure 1.8. The formation of ethene molecule.

In general, in unsaturated compounds the carbon atoms which are held by double bonds are sp^2 hybridised. Thus, in propene, $CH_3CH{=}CH_2$, the two carbon atoms holding the double bond are sp^2 hybridised while the carbon atom of the CH_3- group is sp^3 hybridised.

sp hybridisation

When an *s* orbital and one of the *p* orbitals take part in hybridisation, it is called **sp hybridisation** and the resultant two orbitals are called **sp orbitals**. Their axes are at 180°.

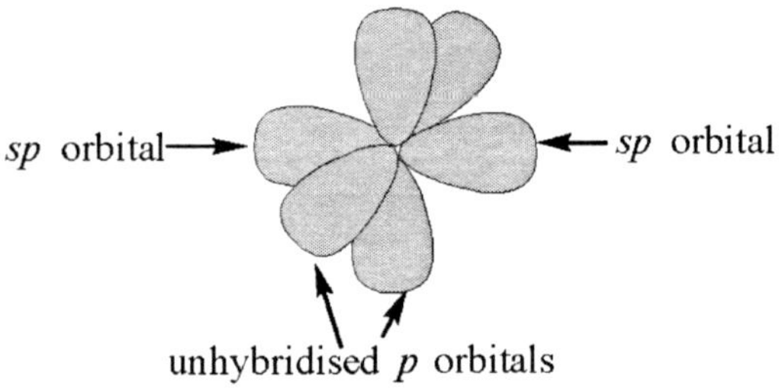

Figure 1.9. *sp* hybrid orbitals of a carbon atom.

Ethyne, C_2H_2, is an unsaturated compound with three covalent bonds (triple bond) between the two carbon atoms. In ethyne, one *sp* orbital of each carbon atom overlaps with a hydrogen orbital each and the other *sp* orbitals of the two carbon atoms overlap endwise, thus forming three σ bonds. The two unhybridised *p* orbitals of each carbon form two π bonds by overlapping sidewise. The electrons in the two pi bonds delocalize to form a cylindrical shaped electron cloud around the C-C sigma bond. The ethyne molecule is linear and the H-C-C bond angle is 180°.

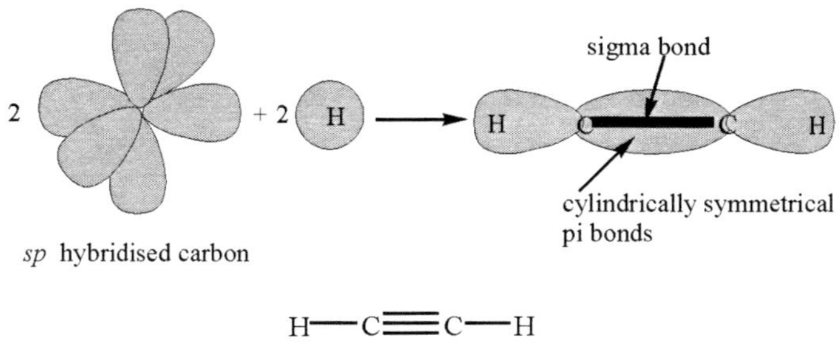

Ethyne

Figure 1.10. The formation of ethyne molecule.

1.5 Electronegativity and polarity of covalent bonds

The covalent bond between two atoms of the same element in a molecule is **non-polar**. This means that the two electrons in the bond are attracted by the two nuclei with the same force. For example, H-H and Cl-Cl bonds are nonpolar. But the covalent bond between two atoms of different elements in a molecule need not be nonpolar. Different atomic nuclei attract the shared electron pair in the bond with different intensity.

Electronegativity in simple terms is a measure of the power of an atom to attract the shared electron pair in a covalent bond. The electronegativity of an element depends mainly on its nuclear charge and atomic size. Electronegativity of elements increases as we go from left to right of the periodic table. It decreases down a group in the periodic table. The electronegativity values of some elements (Pauling's scale) are given in Table 1.2.

Element	*Electronegativity*	*Element*	*Electronegativity*
H	2.2	Mg	1.3
Li	1.0	Al	1.6
Be	1.6	Si	1.9
B	2.0	P	2.2
C	2.5	S	2.6
N	3.0	Cl	3.2
O	3.4	Br	3.0
F	4.0	I	2.7
Na	1.0		

Table 1.2. Electronegativities of some common elements.

A H-Cl bond is polar. Since the chlorine atom is more electronegative than the hydrogen atom, the shared electron pair in hydrogen chloride is attracted more strongly by the chlorine atom, keeping closer to it. This causes the development of a net partial negative charge (δ-) on the Cl end and an equal amount of positive charge (δ+) on the H end of the HCl molecule. Such a bond is called a **polar covalent bond** and has a dipole moment proportional to the electronegativity difference. In the same way, the bonds in molecules like H_2O, NH_3, HFand SO_2 are polar.

δ+ δ-
H—Cl

H⟶Cl
the arrow is used
to indicate the
direction of polarity

Consider the bond between a carbon atom and a chlorine atom. The more electronegative chlorine has a δ- charge and the carbon a δ+ charge. The C-Cl bond is permanently polarized.

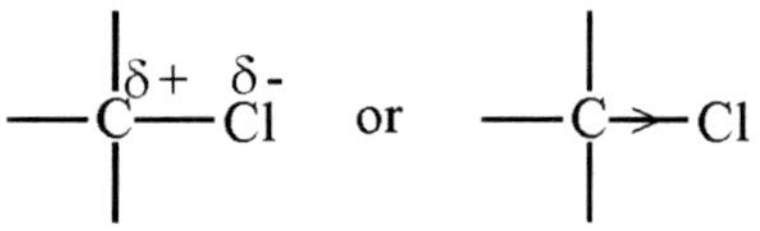

The polarizing effect in single covalent bonds is known as the **inductive effect**. If the $\delta+$ carbon (in the example above) is attached to other carbon atoms, the inductive effect is transmitted along the chain.

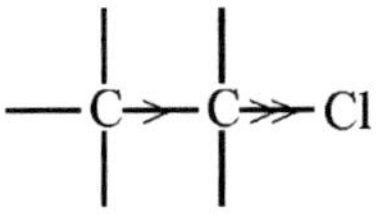

The effect is not very significant beyond the second carbon atom. Halogen atoms, the nitro group (-NO_2) and the phenyl group (C_6H_5-) are examples of groups which have the **electron withdrawing inductive effect**. Other atoms or groups (for example, the methyl group) have the opposite **electron donating (releasing) inductive effect**.

Each covalent bond in a methane molecule is polar but the molecule as a whole is non-polar because of its symmetry. A molecule of CO_2 is non-polar because it is linear, and polarities of the two sets of bonds cancel each other out.

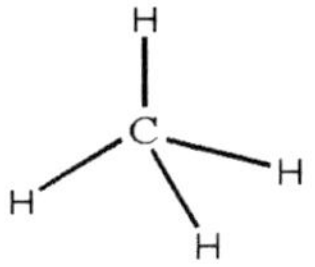

Notes

1. **Valence electrons** are the electrons present in the outermost energy level of an atom.

2. A **lone pair** of electrons is the pair of electrons in an atomic orbital. For example, an oxygen atom ($1s^2 2s^2 2p_x^2 2p_y^1 2p_z^1$) has two lone pairs of electrons in the second energy level which do not take part in bonding and two unpaired electrons which can form two covalent bonds. The two lone pairs of electrons of oxygen in water are shown below.

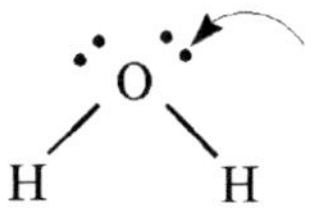

3. Molecules or ions of the type AX_4 (for example, CCl_4 , BF_4^-, NH_4^+) have a tetrahedral shape with A at the centre and X at the corners of the tetrahedron, and all X-A-X bond angles equal to 109.5° approximately.

Tutorial: helping you learn

Practice questions

1. By using dots to represent the valence electrons, write the electronic structure of methanol, CH_3OH.

2. By using dashes to represent covalent bonds, write the structure of methanol, CH_3OH.

3. Indicate the type of hybridisation of the atomic orbitals of each carbon atom in acetonitrile, CH_3CN and give all the bond angles.

Progress questions

1. Write the electronic configuration of the following elements:

   (a) Be (b) B (c) F (d) Ne

   (Atomic number: Be = 4, B = 5, F = 9, Ne = 10)

2. Write the electronic configuration of the elements 12-18 in the periodic table.

3. (a) By using dots to represent the valence electrons, write the electronic structures of the following molecules:

   BCl_3 H_2O CO_2 CH_3Br CH_3CN

   (b) By using dashes to represent the covalent bonds, write the structures of the molecules in (a).

4 Indicate the type of hybridisation of the orbitals of each carbon atom in the following molecules:

   (a) CII_3OH (b) $CH_3CH{=}CH_2$
   (c) $CH_2{=}CH\text{-}CH{=}CH_2$ (d) $ClCH_2CH_2CHO$
   (e) $CH \equiv C{-}CH_3$

5. What is the expected bond angle in each of the following cases?

   (a) H-C-H bond angle in CH_3OH,
   (b) H-C-C, C-C=C and C=C-H bond angles in $CH_3CH{=}CH_2$
   (c) Cl-B-Cl bond angle in BCl_3. (Hint: The boron in BCl_3 is sp^2 hybridised.)

6. Predict the bond angles and the shape of the molecules in each of the following:

   (a) CCl_4 (b) CH_3Br (c) $H_2C{=}O$ (d) $(CH_3)_4N^+$

7. Show the polarity of the bonds given below, using $\delta+$ and $\delta-$ signs (Refer to Table 1.2 for electronegativity values):

   (a) C-O and O-H bonds in CH_3OH (b) C=O bond in $H_2C{=}O$
   (c) C-Cl bond in CH_3Cl (d) H-O bond in H_2O
   (e) N-H bond in NH_3 (f) C-H and C-Br bonds in CH_3Br

Discussion points

The bonds in some molecules are non-polar because of the even sharing of electrons and these are pure covalent bonds. In some compounds, one or more electrons from an element are transferred to atoms of other elements forming ions. Between these two extreme cases of non-polar compounds and ionic compounds, there is a continuous range of compounds with partial covalent and partial ionic character. Discuss the factors affecting these bond types.

Practical assignment

The covalent bond is one of the types of bonding or forces of attraction between atoms in compounds. Make a list of the other types of bonding in substances, intramolecular as well as intermolecular. Describe briefly each of these bonds, with examples.

Study tips

1. It is a good practice to read textbooks, as many as possible and as often as possible.

2. Studying organic chemistry requires reading and remembering a lot of facts. Reading the topic a few times helps to keep it in the memory. The advice of a chemistry teacher many years ago was: 'Read, forget; read, forget and read until you can't forget.'

3. Answering questions and working out problems are important as these help you to realise whether you understand the topic. First, familiarise yourself with the practice questions and then try to answer the progress questions.

2

Organic Compounds

One-minute summary – Hundreds of thousands of organic compounds are known to us, and their study can be complex if not done systematically. They can be grouped according to similarities in structure and properties. Studying the structure and properties of a few compounds is sufficient to predict the nature of other members of that class. Thus, we will discuss classes of compounds such as saturated and unsaturated hydrocarbons, aliphatic and aromatic hydrocarbons, and compounds containing various functional groups. The IUPAC nomenclature of the compounds in each class, and the structure of some typical class members, are given. How to write the structural formula and a condensed version is also shown here. The objectives of this chapter are to:

- name the various classes of organic compounds
- write the systematic names of the compounds in each class
- know the structural and the condensed structural formulae of the compounds
- study structural isomerism
- give a summary of functional groups.

2.1 What are hydrocarbons?

Hydrocarbons are compounds containing carbon and hydrogen only. Methane CH_4, butane C_4H_{10} and benzene C_6H_6 are examples of hydrocarbons. Hydrocarbons can be divided into two main groups; *aliphatic* and *aromatic hydrocarbons*. **Aromatic** hydrocarbons are compounds which contain one or more benzene rings. Other hydrocarbons are termed **aliphatic**. Aliphatic hydrocarbons can be divided into *alkanes, cycloalkanes, alkenes* and *alkynes*. The general formulae, the structures and the names of these classes of compounds are given below.

2.2 Saturated hydrocarbons: alkanes

Alkanes can be considered as the basic class of organic compounds. Alkanes form a homologous series of compounds. A **homologous series** is a series of compounds with similar chemical properties and gradually varying physical properties, all the members of which can be represented by a general formula. Each member in the series differs from the next member by a -CH_2- group. The general formula of alkanes is C_nH_{2n+2}, where n stands for the number of carbon atoms in the molecule. The first four members in the alkane series are given characteristic names while the names of higher members are derived from the Greek or the Latin words for the number. The names of all alkanes end in *-ane*. The names and formulae of the first few members in the series are given below.

Alkane	*Formula*
Methane	CH_4
Ethane	C_2H_6
Propane	C_3H_8
Butane	C_4H_{10}
Pentane	C_5H_{12}
Hexane	C_6H_{14}
Heptane	$C_7 H_{16}$
Octane	C_8H_{18}
Nonane	C_9H_{20}
Decane	$C_{10}H_{22}$

Table 2.1. The formulae of alkanes.

2.3 Drawing chemical structures

We have seen in Chapter 1 that the carbon atom in methane is sp^3 hybridised, the methane molecule is tetrahedral in shape, and each H-C-H bond angle is approximately 109.5°. The carbon atoms in ethane, and other saturated hydrocarbons, are all sp^3 hybridised and they bond tetrahedrally to other atoms. The three-dimensional structures of methane and ethane can be drawn as Figure 2.1.

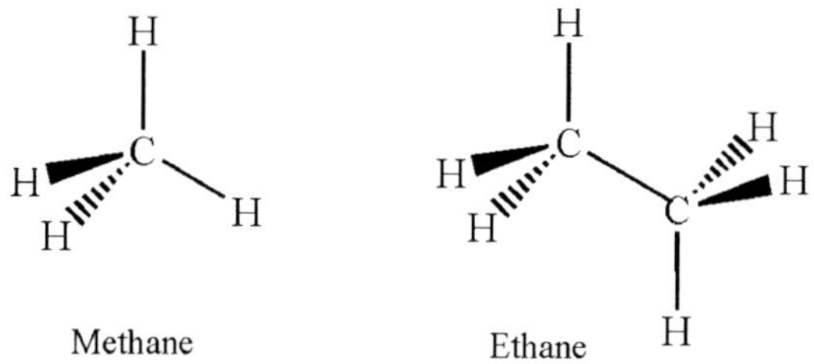

Figure 2.1. Three-dimensional formulae of methane and ethane.

It becomes difficult and time-consuming to draw the three dimensional structures of bigger molecules. A simple way of drawing the structures of methane, ethane and propane is as given below. Remember that all the bond angles, in a tetrahedral arrangement, are about 109.5°.

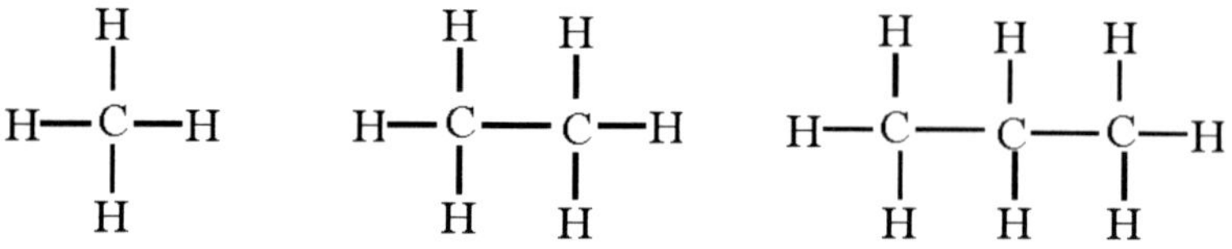

Methane Ethane Propane

The structural formula of a compound is often written in a different way with each carbon atom and the hydrogens attached to that carbon as a group. This way of representation is called the **condensed structural formula**. The condensed structural formulae of methane and ethane are CH_4 and CH_3CH_3 respectively.

There is a yet more simplified way of writing the structural formula and this is called the **skeletal** or **line formula**. The condensed structural formulae and the line formulae of a few compounds are given in Table 2.2 overleaf. Note that in the line formula, each junction represents a CH_2- group, and the end one a CH_3- group. The presence of a substituent on a carbon atom reduces the number of hydrogen atoms by one.

Structural formula	Condensed structural formula	Skeletal formula
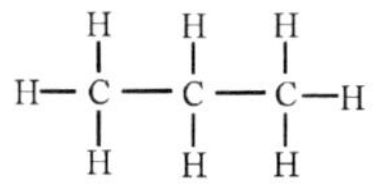	$CH_3\,CH_2\,CH_3$	
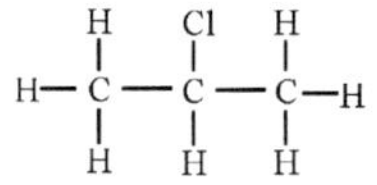	$CH_3\,CHCl\,CH_3$	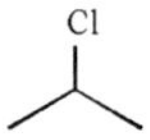
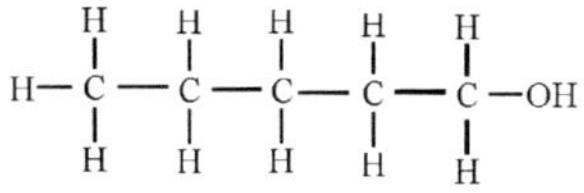	$CH_3\,CH_2\,CH_2\,CH_2\,CH_2OH$	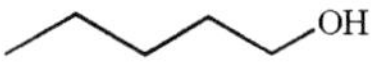

Table 2.2. Different forms of structural formulae.

2.4 The importance of systematic naming

There is an enormous number of organic compounds and it is a difficult task to give them individual names. Though some of the compounds have their common names (trivial names), it is necessary to name the compounds by a set of systematic rules. This set of rules is called the **IUPAC** (International Union of Pure and Applied Chemistry) nomenclature. The IUPAC nomenclature of various classes of compounds is discussed in appropriate places.

2.5 Introducing isomerism

There is only one compound each with the molecular formula CH_4, C_2H_6 and C_3H_8. The structural formulae of these compounds are:

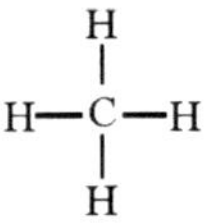
Methane

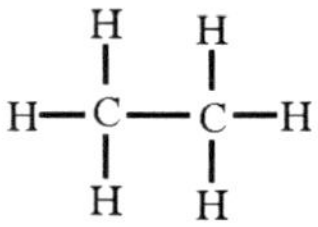
Ethane

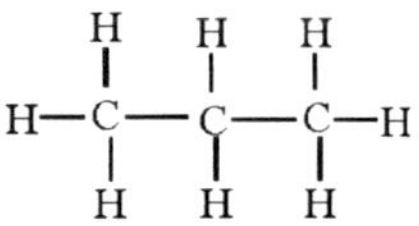
Propane

But we can arrange the carbon and hydrogen atoms of butane, C_4H_{10}, in two different ways as shown in Figure 2.2.

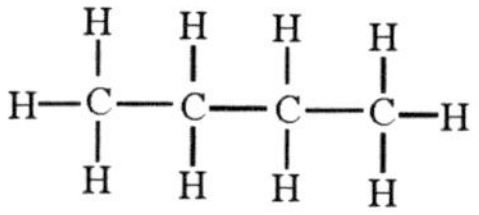

(1)

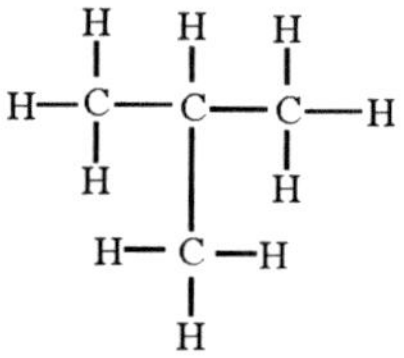

(2)

Figure 2.2. Different arrangements of C_4H_{10}

These are two different compounds with the same molecular formula but different structural formulae and such compounds are called **isomers**. The structures (1) and (2) are isomers of butane. Isomers in which atoms are connected differently are called **constitutional isomers**.

The number of isomers increases as the number of the carbon atoms in the alkane series increases. There are three pentanes, five hexanes, nine heptanes, 75 decanes and so on. Hence the importance of IUPAC nomenclature.

2.6 Alkyl groups

Alkyl groups are structural units derived from alkanes by removing a hydrogen atom. The name of an alkyl group is derived from the corresponding alkane by substituting the -ane ending of the alkane by -yl . Thus the methyl group, CH_3, corresponds to methane, CH_4. The names and formulae of some alkyl groups and their parent alkanes are given below.

Alkane	*Formula*	*Alkyl group*	*Formula*
Methane	CH_4	Mcthyl	CII_3-
Ethane	C_2H_6	Ethyl	CH_3CH_2-
Propane	C_3H_8	Propyl	$CH_3CH_2CH_2$-

Table 2.3. The formulae of some alkanes and alkyl groups.

Nomenclature of alkanes

The straight chain alkanes (alkanes without any branches) are named using the prefix *n*- (normal). Thus $CH_3CH_2CH_2CH_3$ is *n*-butane, $CH_3CH_2CH_2CH_2CH_3$ is *n*-pentane, and $CH_3CH_2CH_2CH_2CH_2CH_2CH_2CH_3$

is *n*-octane. The two isomers of butane are *n*-butane and *iso*-butane.

$CH_3CH_2CH_2CH_3$

n-Butane

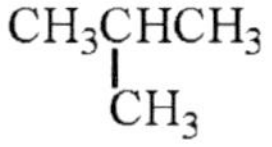

Isobutane

2.7 Naming compounds by the IUPAC system

The rules for naming alkanes by the IUPAC system of nomenclature are given below.

1. The names butane, hexane, octane etc. are used for the unbranched alkanes. The IUPAC name of $CH_3CH_2CH_2CH_2CH_3$ is pentane. Its common name is *n*-pentane.
2. In the case of branched compounds, find the number of carbon atoms in the longest continuous C-C chain; this chain determines the parent name of the compound.

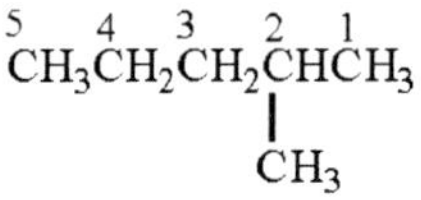

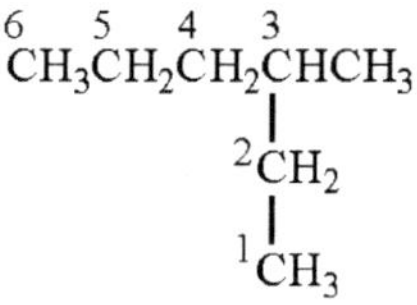

Thus the names of the above compounds are derived from pentane (the five-carbon alkane) and hexane (the six-carbon alkane) respectively.

3. Number the carbon atoms in the longest chain, starting from the end closer to the branch. If there are branches at equal distances from both ends of the main chain, number the carbon atom starting from the end closer to the second branch and so on. Note the following examples:

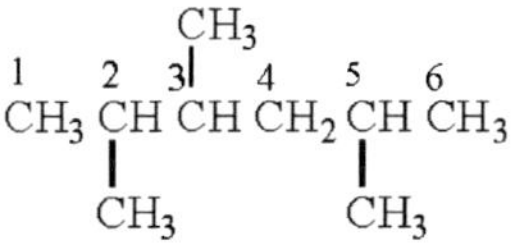

4. Name the substituent, assign a number to locate it in the main chain and give the name as shown below.

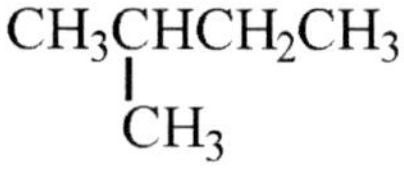

2-Methylbutane

In the example given, the parent alkane is butane and there is a methyl group attached to C2 of butane.

5. When there are two or more substituents, give each substituent a number to locate it and list the substituents alphabetically. When two or more identical substituents are present, use the prefix *di-*, *tri-*, *tetra-* etc. with the name. These prefixes *di-*, *tri-* and *tetra-* are ignored when arranging the substituents alphabetically. When there are two substituents attached to the same carbon, use the number to locate each substituent. Note the following examples:

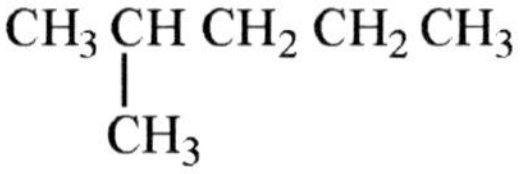

2-Methylpentane

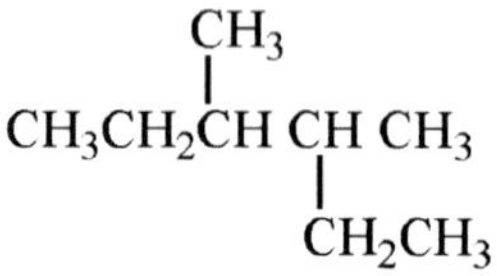

3,4-Dimethylhexane

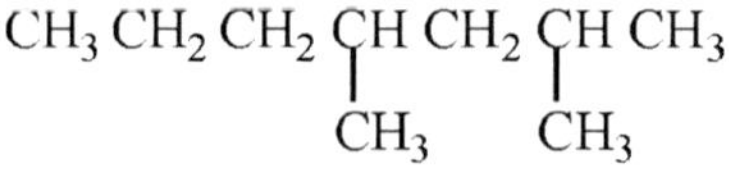

2,4-Dimethylheptane

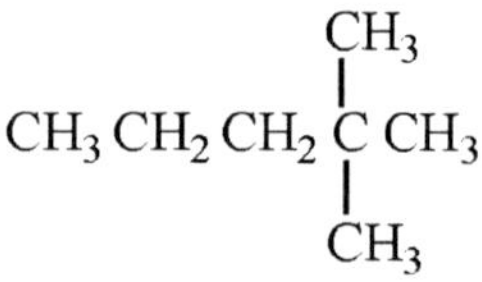

2,2-Dimethylpentane

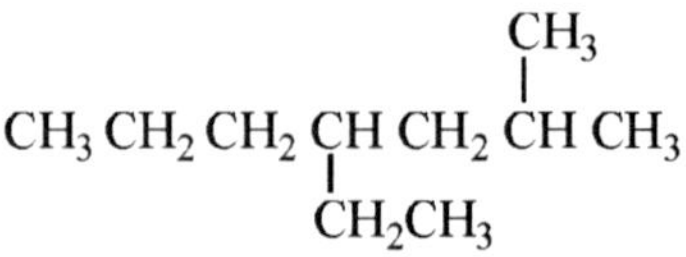

4-Ethyl-2-methylheptane

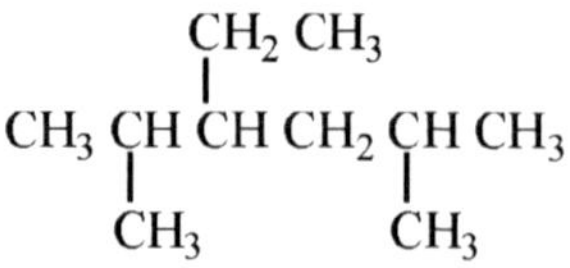

3-Ethyl-2,5-dimethylhexane

2.8 Cycloalkanes

Cycloalkanes are saturated, closed chain compounds which have a C-C ring and are represented by the general formula C_nH_{2n}. The first member of the series is cyclopropane, C_3H_6. This is followed by cyclobutane, C_4H_8; cyclopentane, C_5H_{10}; cyclohexane, C_6H_{12} and so on.

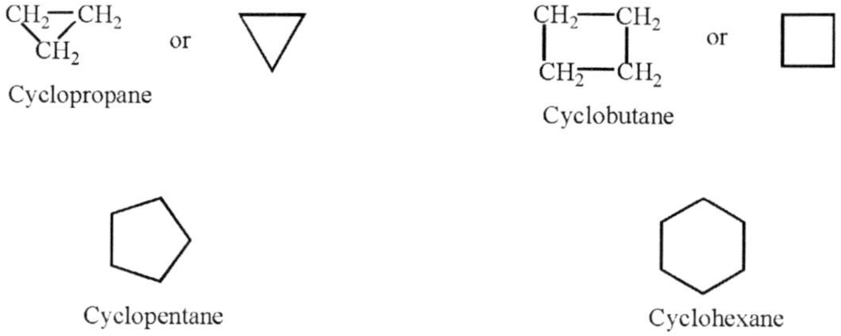

Figure 2.3. Some cycloalkanes.

2.9 Unsaturated hydrocarbons: alkenes

Alkenes are unsaturated hydrocarbons containing a double bond in the molecule. Alkenes form a homologous series and the general formula of alkenes is C_nH_{2n}. In the IUPAC nomenclature, the names of alkenes are derived from the corresponding alkanes by substituting -ene for -ane. The names and formulae of the first few members in the series are given below.

Alkane	*Formula*	*Alkene*	*Formula*
Methane	CH_4		
Ethane	C_2H_6	Ethene	C_2H_4
Propane	C_3H_8	Propene	C_3H_6
Butane	C_4H_{10}	Butene	C_4H_8
Pentane	C_5H_{12}	Pentene	C_5H_{10}

Table 2.4. The formulae of some alkenes.

There is only one compound (no isomers) each of ethene and propene. The structural formulae of ethene and propene are:

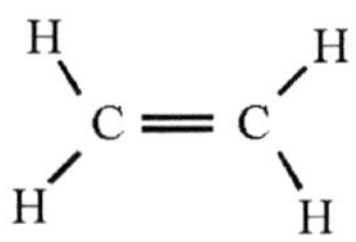

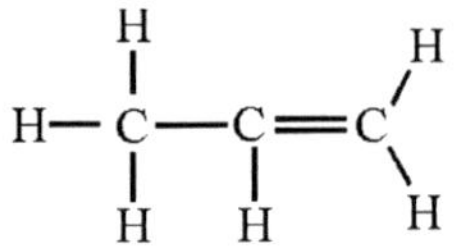

$CH_2=CH_2$ $CH_3\,CH=CH_3$

Ethene Propene

The atoms in butene, C_4H_8, can be arranged in different ways as shown below. These are constitutional isomers of butene. The structures (1) and (2) are straight chain alkenes which differ in the position of the double bond and (3) is a branched chain alkene.

$CH_2=CH\,CH_2\,CH_3$ (1)

$CH_3\,CH=CH\,CH_3$ (2)

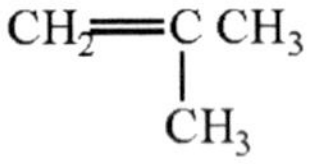

(3)

IUPAC rules for naming alkenes

1. Find the longest C-C chain which contains the double bond. The parent name is given on the basis of this chain. For example, structure (4) is named from butene, the four carbon alkene and (5) from pentene, the five carbon alkene.

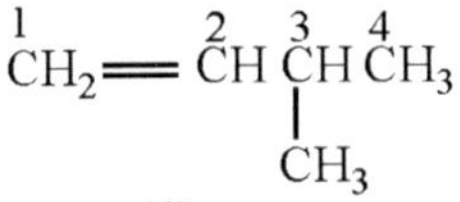

(4)

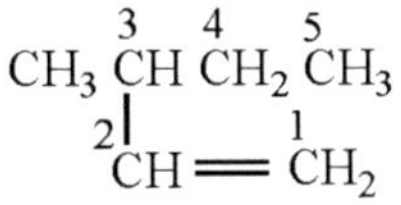

(5)

2. Number the carbon atoms in the chain starting from the end closer to the double bond. If the double bond is placed at equal distances from both ends, then start numbering the chain closer to a branch and so on. These points are exemplified in structures (4), (5) and (6).

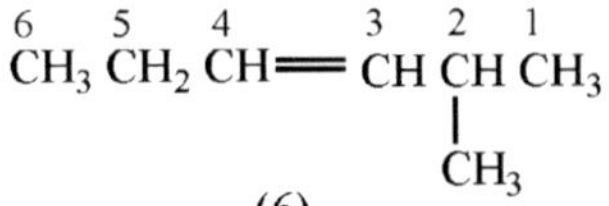

(6)

3. To assign the position of the double bond, use the lower of the two numbers of the carbon atoms which hold the double bond as prefix or suffix, as shown below.

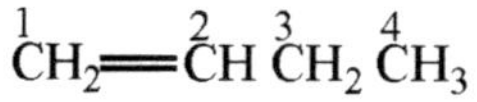

1-Butene or but-1-ene

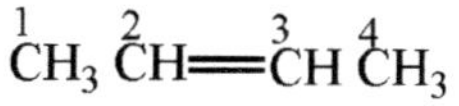

2-Butene or but-2-ene

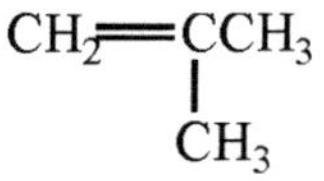

2-Methylpropene

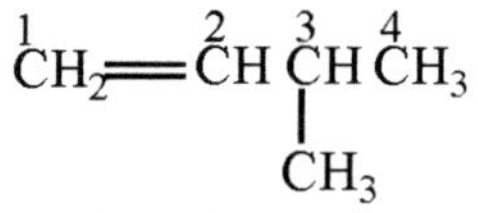

3-Methylbut-1-ene

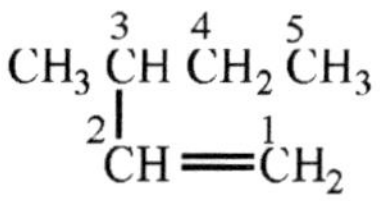

3-Methylpent-1-ene

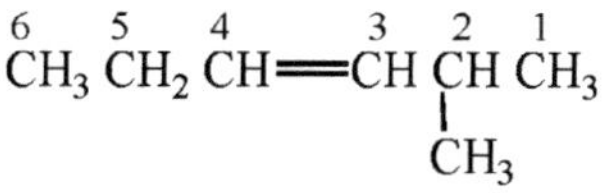

2-Methylhex-3-ene

2.10 Unsaturated hydrocarbons: alkynes

Alkynes are unsaturated hydrocarbons with a triple bond in the molecule. The members can be represented by the general formula C_nH_{2n-2} and the names of alkynes end in -yne. The first member in the series is ethyne (acetylene), C_2H_2. The names and formulae of the first few members of the alkynes are given below.

Alkane	*Formula*	*Alkyne*	*Formula*
Methane	CH_4		
Ethane	C_2H_6	Ethyne	C_2H_2
Propane	C_3H_8	Propyne	C_3H_4
Butane	C_4H_{10}	Butyne	C_4H_6

Table 2.5. The formulae of some alkynes.

Alkynes can be named by applying similar rules as in the case of alkenes. Note the examples given below:

3-Methylbut-1-yne

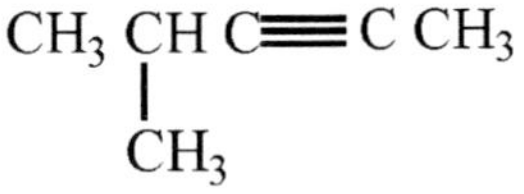

4-Methylpent-2-yne

2.11 Aromatic hydrocarbons

Aromatic hydrocarbons contain one or more benzene rings. Benzene, C_6H_6, is the basic aromatic hydrocarbon. A benzene molecule has six carbon atoms which form a ring. All the six carbon atoms are sp^2 hybridised. An sp^2 hybridised carbon atom has three sp^2 orbitals containing an electron each. These orbitals are in one plane and are directed towards the corners of an equilateral triangle. The two lobes of the unhybridised p orbital which hold one electron project above and below the plane of the sp^2 orbitals (Section 1.4).

Each carbon atom uses two sp^2 orbitals to overlap with similar orbitals of two neighbouring carbon atoms and the third sp^2 with the s orbital of a hydrogen atom, forming σ covalent bonds. All the carbon atoms and hydrogen atoms are in one plane. The six electrons in the unhybridised p orbitals delocalise to form an electron cloud located above and below the ring. A simplified representation of this is given by structure (3) below. The ring stands for the six p electron cloud. For simplicity, we can consider that these electrons form three double bonds which are located at alternate positions in the ring as in (1).

The formula of benzene is often written as (1), with alternating single and double bonds, although all six carbon-carbon bonds are identical with a bond length between those of a double bond and a single bond. The skeletal formula of benzene is written as (2) or (3).

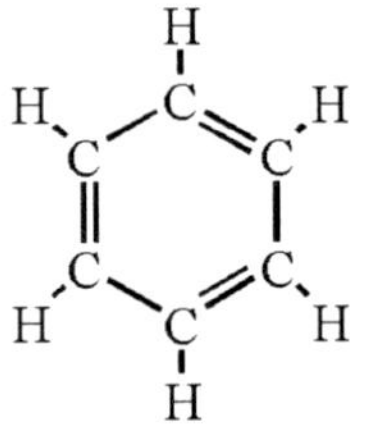

(1)

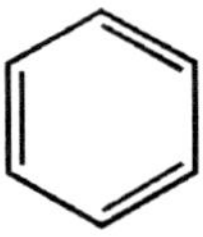

(2)

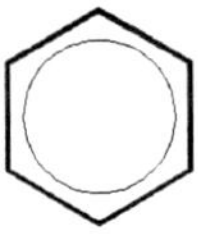

(3)

Benzene

▶ *Note* – A typical C-C bond length is 0.154 nm and C=C bond length, 0.134 nm. All the carbon-carbon bond lengths in benzene are identical and the value is 0.139 nm.

When one or more hydrogen atoms of benzene are replaced by other atoms or groups of atoms, substituted aromatic compounds are formed. When there is more than one substituent, they are located by numbers. The names of some substituted benzenes are given below. Many aromatic compounds have trivial names which are often used. The trivial names are given in brackets.

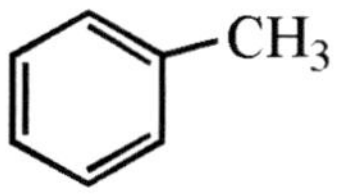

Methylbenzene
(Toluene)

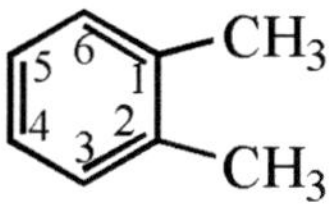

1,2-Dimethylbenzene
(o-Xylene)

▶ *Note* – Disubstituted benzenes also use the prefixes *ortho*- (*o*-) for 1,2- position, *meta*- (*m*-) for 1,3- position and *para*- (*p*-) for 1,4- position.

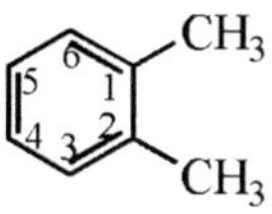

1,2-Dimethylbenzene
(o-Xylene)

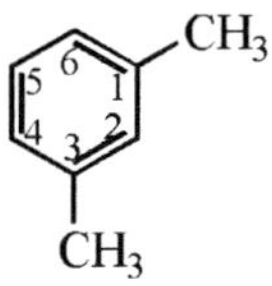

1,3-Dimethylbenzene
(m-Xylene)

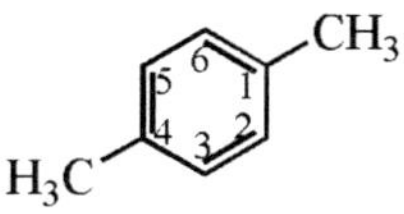

1,4-Dimethylbenzene
(p-Xylene)

2.12 What are functional groups?

When one or more hydrogen atoms of an alkane are substituted by atoms or groups of atoms containing oxygen, sulphur, nitrogen, halogens etc. we get classes of compounds, whose properties mainly depend on the substituents. These substituents are generally called **functional groups**. Double and triple bonds are also considered as functional groups. Some classes of compounds, their functional groups and typical examples are given below.

2.13 Alcohols

When one or more hydrogen atoms of an alkane are substituted by hydroxyl, -OH, groups, alcohols are formed. Monohydric alcohols contain one hydroxyl group. They form a homologous series of compounds with the general formula $C_nH_{2n+1}OH$. The names of alcohols are derived from the corresponding alkanes by substituting the *-e* ending of an alkane by *-ol*. The first three members are methanol (CH_3OH), ethanol (C_2H_5OH) and propanol (C_3H_7OH).

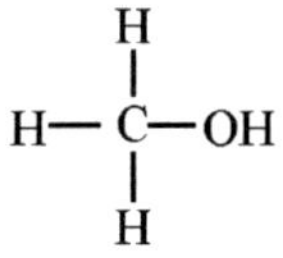

Methanol

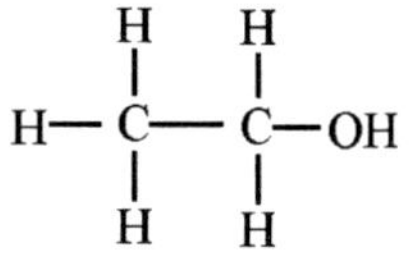

Ethanol

Propanol has two constitutional isomers: (1) and (2). The trivial names of these isomers are n-propanol and iso-propanol respectively.

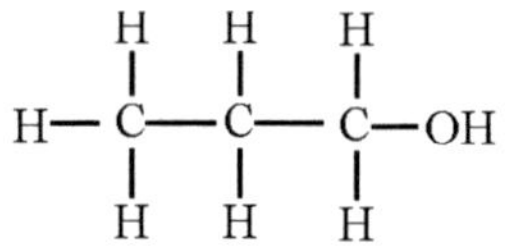

(1)

n-Propanol

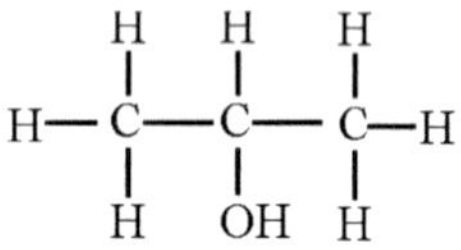

(2)

Isopropanol

The summary of the rules for the IUPAC nomenclature of alcohols arc:

1. The longest C-C chain containing the hydroxyl group is numbered, starting from the end closer to the -OH group.
2. The location of the hydroxyl group is shown by a number.
3. The other substituents are named accordingly. The systematic names of the isomers of propanol are 1-propanol and 2-propanol.

$\overset{3}{C}H_3\overset{2}{C}H_2\overset{1}{C}H_2OH$

1-Propanol

Propan-1-ol

$\overset{1}{C}H_3\overset{2}{C}HOH\overset{3}{C}H_3$

2-Propanol

Propan-2-ol

2.14 Phenols

Phenols are compounds containing a hydroxy group or groups attached to a benzene ring. The simplest phenol has the molecular formula C_6H_5OH. It is called hydroxybenzene or phenol. A few examples of phenols are given below.

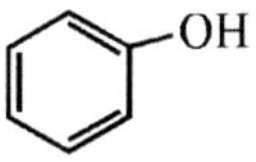

Phenol

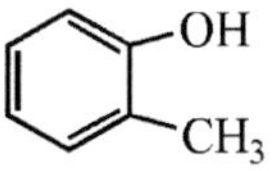

2-Methylphenol

(o-Cresol)

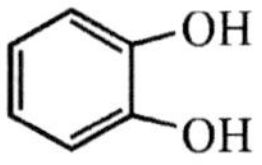

1,2-Benzenediol

(Catechol)

2.15 Aldehydes and ketones

Aldehydes are compounds which contain a -CHO group, and ketones a -CO group. Both aldehydes and ketones contain a carbonyl group, -C=O. In aldehydes the carbonyl carbon is bonded to a hydrogen atom and to an alkyl or aryl group (or to another hydrogen atom). In ketones the carbonyl carbon is connected to two alkyl or aryl groups.

Aldehydes form a homologous series of compounds, the names of which are derived from the corresponding alkanes, by substituting the *-ane* ending of alkane with *-al*. The aldehyde group is always at the end of a chain and the aldehyde carbon is C-1. The other substituents are named accordingly.

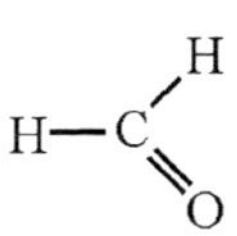

HCHO

Methanal

(Formaldehyde)

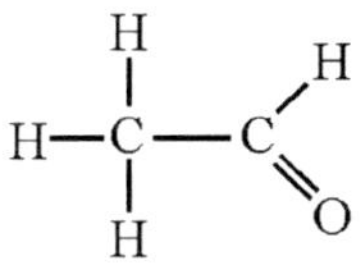

CH_3 CHO

Ethanal

(Acetaldehyde)

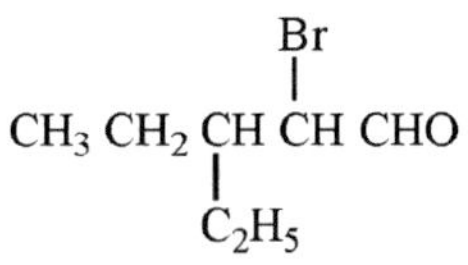

2-Bromo-3-ethylpentanal

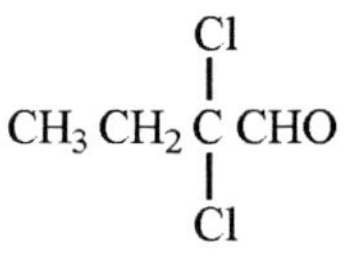

2,2-Dichlorobutanal

The smallest ketone has three carbon atoms. The names of ketones end in *-one*. The carbon chain is numbered starting from the end closer to the carbonyl group.

$$CH_3\underset{\underset{O}{\|}}{C}CH_3 \quad \text{or} \quad CH_3\,CO\,CH_3$$

Propanone
(Acetone)

$CCl_3\,CO\,CH_3$

1,1,1-Trichloropropanone

$CH_3\,CO\,CH_2\,CH_2\,CH_3$

2-Pentanone or pentan-2-one

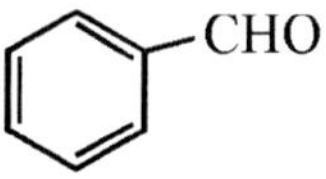

Benzaldehyde

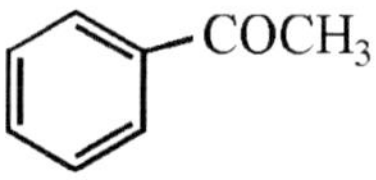

Methylphenylketone
(Acetophenone)

2.16 Carboxylic Acids

These compounds contain the -COOH (-CO_2H) group. Monocarboxylic acids contain one -COOH group. The names of carboxylic acids are derived from the corresponding alkane by replacing the *-e* ending of the alkane with *-oic acid*. Thus the acid with one carbon atom is methanoic acid, with two carbon atoms, ethanoic acid and so on. The carboxyl carbon is C-1 in a chain.

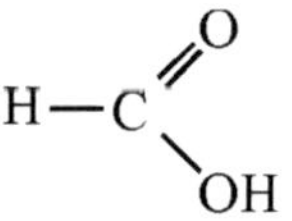

or HCOOH

Methanoic acid
(Formic acid)

$CH_3\,COOH$

Ethanoic acid
(Acetic acid)

$$CH_3\underset{\underset{CH_3}{|}}{C}Cl\,COOH$$

2-Chloro-2-methyl
propanoic acid

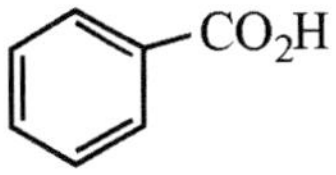

Benzoic acid

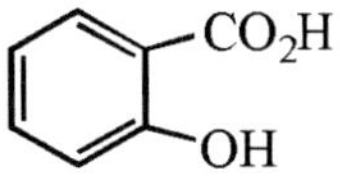

2-Hydroxybenzoic acid
(Salicylic acid)

2.17 Amines

Amines are organic bases and they can be considered as compounds where one or more hydrogen atoms of ammonia are substituted by alkyl or aryl groups. If one hydrogen atom of ammonia is substituted by an alkyl group, a primary amine is formed. Replacing two hydrogens by two alkyl groups gives a secondary amine, and all three hydrogens by three alkyl groups, a tertiary amine. Amines are named by adding -*amine* to the name of the alkyl or aryl group/s. Note the names of the following compounds:

CH_3NH_2

Methylamine

CH_3NHCH_3

Dimethylamine

$CH_3CH_2CH_2NH_2$

1-Propylamine

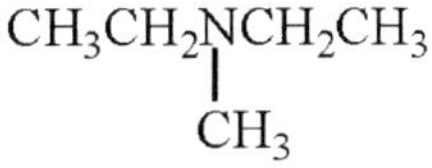

Diethylmethylamine

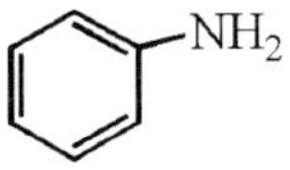

Phenylamine
(Aniline)

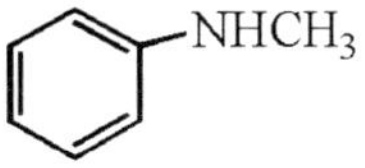

N-Methylphenylamine

2.18 Summary of functional groups

Various classes of compounds and the functional groups are summarised below.

Class or Family	Functional group	Example
Alkenes (double bond)	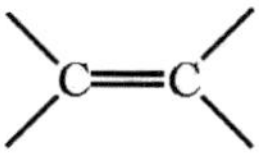	$CH_3CH = CH_2$ propene
Alkynes (triple bond)	$-C\equiv C-$	$CH_3C\equiv CH$ propyne
Alcohols	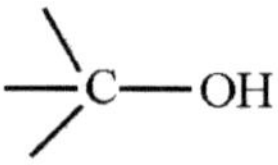	CH_3CH_2OH ethanol
Ethers	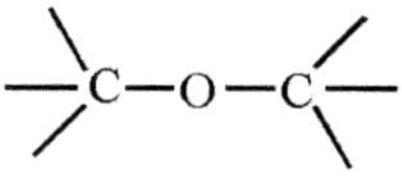	CH_3OCH_3 dimethylether
Aldehydes	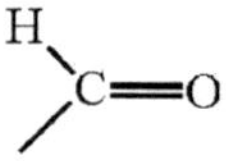	CH_3CHO ethanal
Ketones	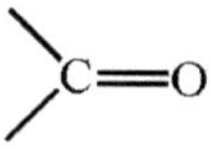	CH_3COCH_3 propanone
Carboxylic acids	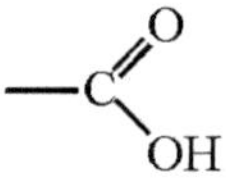	CH_3COOH ethanoic acid
Esters	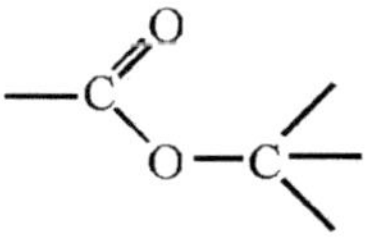	CH_3COOCH_3 methyl ethanoate
Carboxylic acid chlorides	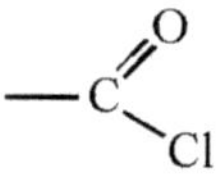	CH_3COCl ethanoyl chloride
Amines	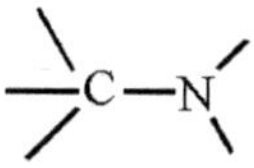	CH_3NH_2 methylamine

Table 2.6. Functional groups.

Tutorial: helping you learn

Practice questions

1. Give IUPAC names for the following compounds:

(a) 

(b) 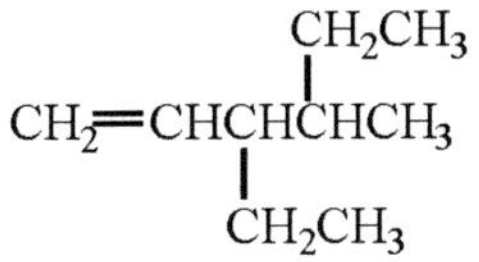

(c) 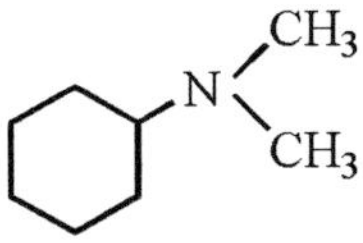

(d) 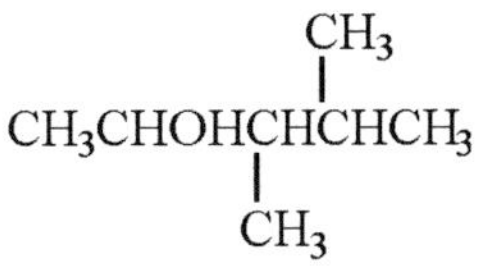

Question hints and tips

(i) In compound (a), select the C-C chain containing the -COOH group. C-1 is the carbon in the COOH group. Number the carbon atoms in the chain so that there are more substituents in the main chain; in this case an ethyl group on C-2 and a chlorine on C-3.

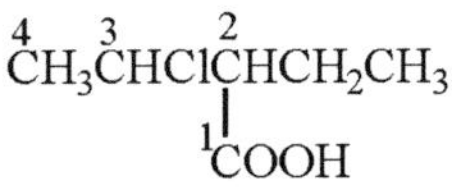

Not

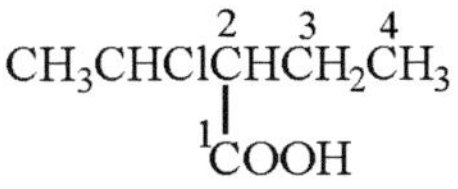

(ii) In compound (b), the longest C-C chain containing the double bond contains six carbons and is thus named from hexene.

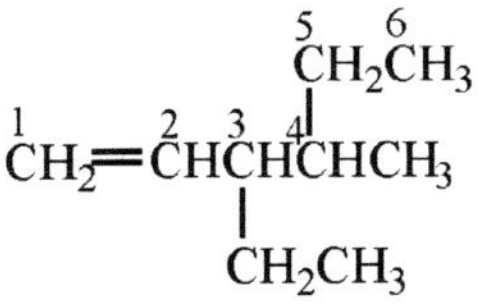

Not

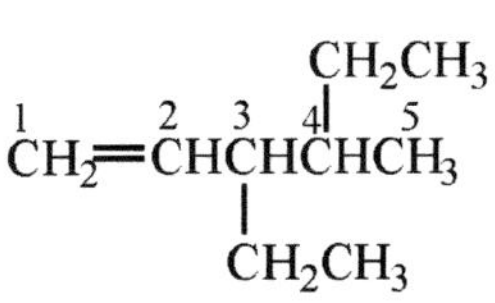

2. Write the skeletal formulae of the compounds in Question 1.

Progress questions

1. Name the functional group/s in each of the following compounds.

(a) $CH_2{=}CHCOOH$ (b) $CH_3CH_2CH_2CH_2OH$

(c) $C_6H_5CH{=}CH_2$ (d) $CH_3CH_2N(CH_3)_2$

(e) C_6H_5CHO (f) $CH_3CH_2COCH_2CH_3$

2. Write the IUPAC names of all the compounds given in Question 1.

3. Write the condensed structural and skeletal formulae of the following compounds:

   (a) 2-Pentyne (b) 2,3-Dichlorobutanoic acid

   (c) 2,3-Dimethyl-2-butanol (d) 1,3,5-Trimethylbenzene

   (e) 3-Hydroxyheptanoic acid (f) 4-Ethyl-3-methylhexanal

4. Give the condensed structural formulae and the IUPAC names of the five isomers of hexane, C_6H_{14}.

Discussion points

Many hydrocarbons are used on a large scale for domestic and industrial purposes. List and discuss the uses of some important hydrocarbons.

Practical assignment

As we have discussed earlier in this chapter, constitutional isomers are compounds with the same molecular formula but with structures in which atoms are connected differently. Thus, butane is isomeric with methylpropane, 1-pentene with 2-pentene, 1-butyne with 1,3-butadiene, ethanol with dimethylether, propanal with propanone, and so on. Make a list of isomers with different functional groups and same molecular formula.

Study tips

1. To get to grips with writing condensed structural and skeletal formulae of compounds, first practise writing the elaborate structural formula with the main -C-C- skeleton in a straight line, then draw two dashes radiating out from each carbon, one above and one below. Each carbon atom should now have four dashes surrounding it. A hydrogen atom or other substituent is then placed at the end of each dash. For example, to write the structural formula of 2,3-dichloropentane, first draw a skeleton with five carbon atoms:

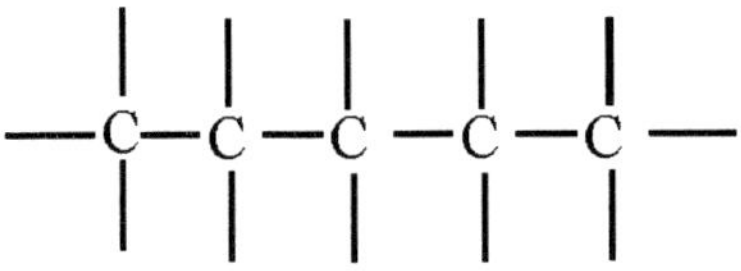

then add the two chlorine substituents onto C2 and C3; and hydrogen atoms onto the rest of the dashes.

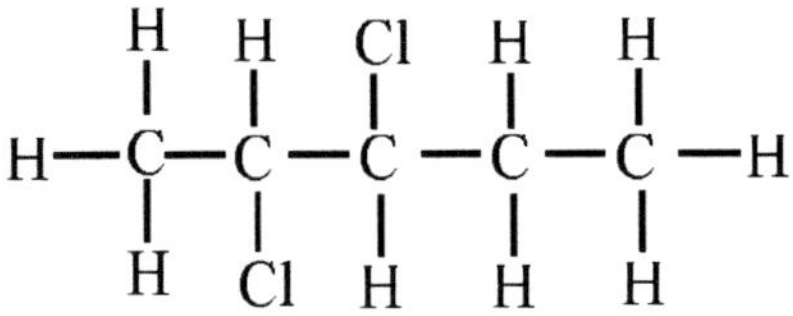

2. Count the number of hydrogen atoms on the end carbon atoms and the carbon atoms inside the chain. Note that the end carbon atoms are bonded to three hydrogen atoms and a carbon inside a chain to two hydrogen atoms. A carbon which carries substituents has one less hydrogen for every substituent atom and so on. The condensed structural formula for the above compound is:

$$CH_3CHClCHClCH_2CH_3$$

3

Alkanes and Halogenoalkanes

One-minute summary – Alkanes are saturated, open-chained hydrocarbons. They are not very reactive compounds. Two common types of reactions that alkanes undergo are substitution and oxidation. Halogenoalkanes are compounds formed when one or more hydrogen atoms of an alkane are substituted by halogen atoms. They are obtained by substituting, hydrogen atoms of an alkane or the hydroxyl group of an alcohol by halogen atoms under suitable conditions. Halogenoalkanes undergo substitution and elimination reactions. The mechanism of reactions, the 'reaction pathway' is introduced in this chapter. This chapter summarises:

- the properties of alkanes
- oxidation/combustion reactions of alkanes
- substitution reactions of alkanes
- the mechanism of substitution reactions
- methods of preparation of halogenoalkanes
- reactions of halogenoalkanes.

3.1 Alkanes

Alkanes form an important class of compounds since they form the basic framework for other classes of compounds. The general formula of alkanes is C_nH_{2n+2} where n stands for the number of carbon atoms in a molecule. The names and formulae of some alkanes (Section 2.2), their IUPAC nomenclature (Section 2.7) and constitutional isomerism (Section 2.5) are discussed in Chapter 2. Many alkanes are used as solvents and fuels and as raw materials for the preparation of other organic compounds.

3.2 The properties of alkanes

The first four alkanes are gases, the higher members being liquids or waxy solids. The melting and boiling points of *n*-alkanes increase gradually with the increase in molecular weight. This is because the strength of intermolecular forces of attraction increases with molecular weight.

The main type of intermolecular forces of attraction in alkanes are **van der Waals forces**. This is the attraction between **instantaneous dipoles**. Since electron clouds around nuclei in molecules are not rigid and can fluctuate, a non-polar molecule can become a dipole at any moment when the centre of positive charge does not coincide with the centre of negative charge. This instantaneous dipole polarizes a neighbouring molecule and a force of attraction is developed between them. Instantaneous dipoles are formed and 'die' continuously. The strength of van der Waals forces increases with electron density which in turn depends on molecular weight. Note the gradual increase in boiling points with the increase in molecular weight (Table 3.1).

An unbranched alkane has a higher boiling point than a branched alkane of the same molecular weight. Linear molecules can pack closer and so the intermolecular forces are stronger. The boiling point of *n*-hexane is 69°C, whereas the boiling point of its isomer 2-methylpentane is 60°C.

Alkane	*Formula*	*Molecular weight*	*Boiling point (°C)*
Methane	CH_4	16	-164
Ethane	C_2H_6	30	-88
Propane	C_3H_8	44	-42
n-Butane	C_4H_{10}	58	0
n-Pentane	C_5H_{12}	72	36
n-Hexane	C_6H_{14}	86	69
2-Methylpentane	C_6H_{14}	86	60

Table 3.1. The properties of alkanes.

Alkanes are insoluble in water but are soluble in or miscible with halogenoalkanes, ethers and some other organic solvents. While alkane molecules are bonded by van der Waals forces, water molecules are held together by stronger hydrogen bonds. Since alkanes cannot form any strong bonds with water molecules, or break the bonds between water molecules, they do not mix. But alkane molecules can develop van der Waals forces with other organic solvents, hence the miscibility.

3.3 Reactions of alkanes

Alkanes are generally not very reactive under normal reaction conditions. This is because of the strong sigma bonds in the molecules (Section 1.3) and also because they are generally non-polar compounds. However, they undergo combustion and substitution reactions.

Combustion or oxidation reactions

Alkanes burn in air or oxygen to form carbon dioxide and water. These are highly exothermic reactions, hence the use of some alkanes as fuels.

$$CH_4\,(g) + 2\,O_2\,(g) \longrightarrow CO_2\,(g) + 2\,H_2O\,(l), \quad \Delta H^0_c = -890.3 \text{ kJ mol}^{-1}$$

$$C_3H_8\,(g) + 5\,O_2\,(g) \longrightarrow 3CO_2\,(g) + 4H_2O\,(l), \quad \Delta H^0_c = -2219.2 \text{ kJ mol}^{-1}$$

► *Note* – The standard heat of combustion ($\Delta H^\circ{}_c$) is the heat evolved when one mole of a compound is burnt completely in air or oxygen.

Substitution reactions: halogenation

1. *With chlorine*

If a mixture of methane and chlorine is heated or exposed to ultraviolet or visible light, they react to form chloromethane and hydrogen chloride.

$$CH_4\,(g) + Cl_2\,(g) \longrightarrow \underset{\text{Chloromethane}}{CH_3Cl\,(g)} + HCl\,(g)$$

In this reaction, a hydrogen atom of methane is substituted by a chlorine atom. Such a reaction is called a **substitution reaction**. Since chlorine is the substituent in this reaction, this type of reaction is called a **chlorination reaction**.

If excess chlorine is present in the reaction mixture, all the hydrogen atoms can be substituted one after the other according to the following equations:

$$CH_3Cl + Cl_2 \longrightarrow \underset{\text{Dichloromethane}}{CH_2Cl_2} + HCl$$

$$CH_2Cl_2 + Cl_2 \longrightarrow \underset{\substack{\text{Trichloromethane} \\ \text{(Chloroform)}}}{CHCl_3} + HCl$$

$$CHCl_3 + Cl_2 \longrightarrow \underset{\substack{\text{Tetrachloromethane} \\ \text{(Carbon tetrachloride)}}}{CCl_4} + HCl$$

The reaction mechanism (Reaction pathway) of halogenation

The reaction mechanism between methane and chlorine can be explained as follows. If a mixture of methane and chlorine is left in the dark, there is no reaction. In the presence of light or heat, a few chlorine molecules absorb energy and dissociate into atoms.

$$Cl{-}Cl \longrightarrow Cl\bullet + Cl\bullet \qquad (1)$$

► *Remember* – A covalent bond is a shared electron pair between two atoms and when it breaks equally, an electron goes to each atom.

The chlorine atom is a very reactive species because it has seven electrons in the outer orbit and has a great tendency to acquire one more electron. It therefore abstracts a hydrogen atom from methane to form hydrogen chloride. A methyl radical is produced simultaneously.

Initiation step

Methyl radical

(2)

The methyl radical is very reactive due to the presence of a single electron on the carbon atom, and it thus combines with another chlorine molecule or a chlorine atom to form chloromethane.

Propagation step

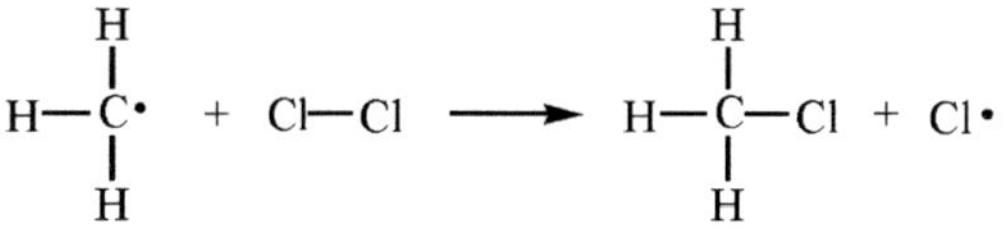

(3)

Termination step

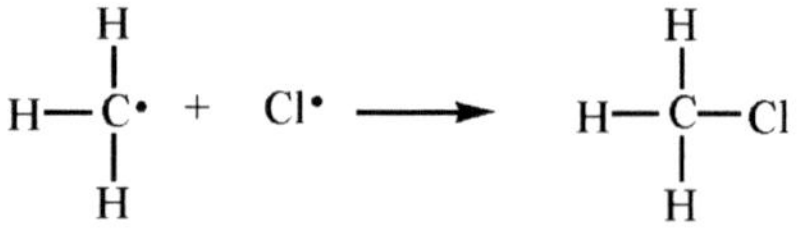

(4)

A chain reaction sets in as the chlorine atoms produced in the third step (Equation 3) react with more methane molecules. This type of substitution reaction is called free radical substitution since free radicals take part in the substitution reaction.

2. With bromine

When propane is heated with bromine, a mixture of 1-bromopropane and 2-bromopropane is formed together with hydrogen bromide.

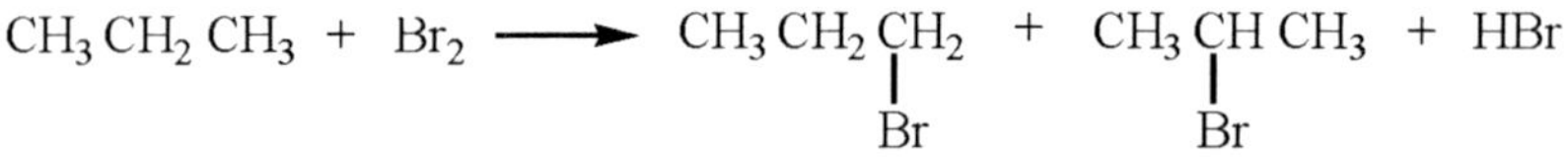

1-Bromopropane 2-Bromopropane

With this reaction, the yield of 2-bromopropane is much higher (92%) than that of 1-bromopropane (8%).

Similarly, bromination of 2-methylbutane produces 2-bromo-2-methyl-butane only.

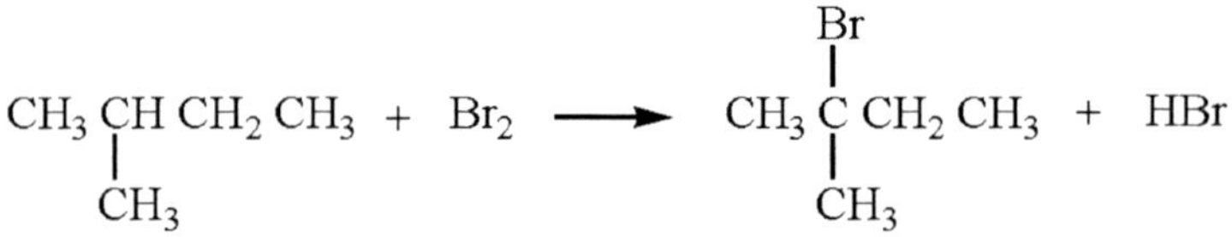

2-Methylbutane 2-Bromo-2-methyl butane

These two reactions demonstrate that a tertiary carbon is more reactive than a secondary carbon, which in turn is more reactive than a primary carbon.

▶ *Note* – A primary carbon (1°) has one alkyl substituent attached to it, a secondary carbon (2°), two alkyl substituents and a tertiary carbon (3°), three alkyl substituents.

Example

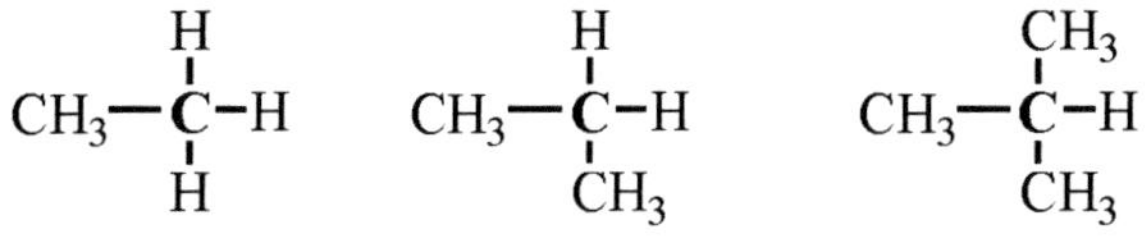

A primary carbon A secondary carbon A tertiary carbon

3. *With fluorine and iodine*

Fluorine reacts with alkanes vigorously and exothermically and such reactions are difficult to control. This is because fluorine is a powerful oxidising agent. On the other hand, iodine does not react with alkanes easily.

4. *Side chain halogenation of toluene*

When an equimolar (equal number of moles) mixture of toluene and chlorine is allowed to react in the presence of UV light, benzyl chloride and hydrogen chloride are formed. Here, a hydrogen atom of the methyl group is substituted and not the hydrogens of the aromatic ring.

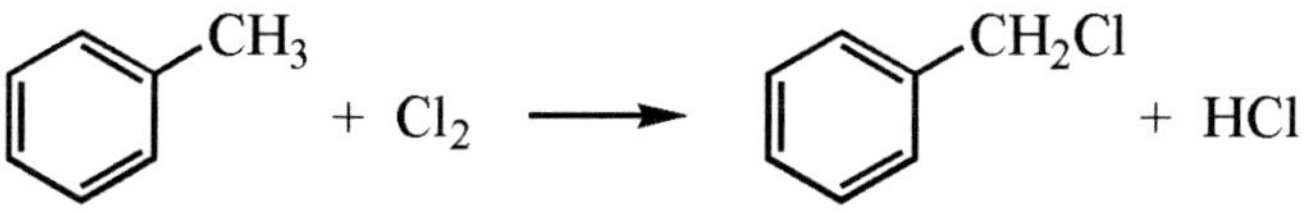

Toluene Benzyl chloride

The chlorine atom abstracts a methyl hydrogen from toluene to form a benzyl radical. The benzyl radical is quite stable as the odd electron can delocalize into the electron cloud system of the aromatic ring (Section 6.7). It then combines with a chlorine atom or a chlorine molecule, according to the free radical mechanism.

With excess chlorine, the other two hydrogen atoms of the methyl group are successively substituted to form benzal chloride, $C_6H_5CHCl_2$ and benzotrichloride, $C_6H_5CCl_3$ respectively.

3.4 Halogenoalkanes

When one or more hydrogen atoms of an alkane are substituted by halogen atoms, halogenoalkanes are formed. Halogenoalkanes are also called **alkyl halides**. The names of halogenoalkanes can be obtained by adding the prefix fluoro-, chloro-, bromo and iodo- to the name of the parent compound or by adding the suffix fluoride, chloride, bromide and iodide to the name of the alkyl group. Thus CH_3Cl is chloromethane or methyl chloride and CF_2Cl_2 is dichlorodifluoromethane. The position of the halogen in the molecule is noted by numbers, remembering that the C-C chain is numbered starting from the end closer to the halogen or an alkyl substituent, giving equal precedence to both halogen and alkyl group.

$CH_3\,CH_2\,CH_2I$

1-Iodopropane

$CH_3\,CCl_2\,CH_3$

2,2-Dichloropropane

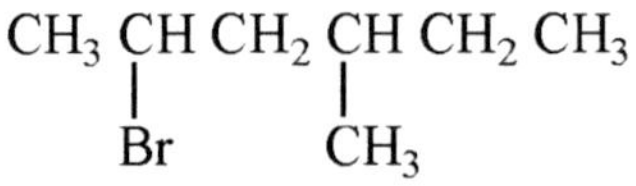

2-Bromo-4-methylhexane

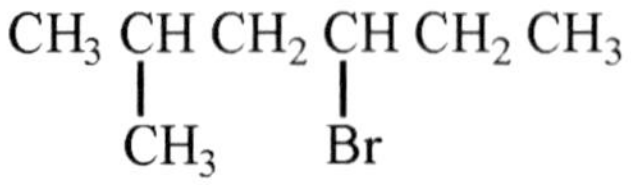

4-Bromo-2-methylhexane

Halogenoalkanes can be classified as primary, secondary and tertiary. In a primary halogenoalkane, the halogen is bonded to a carbon which is bonded to one alkyl group and two hydrogen atoms at least. In a secondary onc, the halogen is attached to a carbon atom which is bonded to one hydrogen and two alkyl groups; and in a tertiary, to a carbon which is bonded to three alkyl groups.

CH_3CH_2Br

Bromoethane
(Primary)

$CH_3CHBrCH_3$

2-Bromopropane
(Secondary)

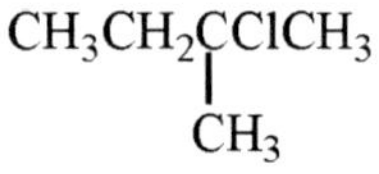

2-Chloro-2-methylbutane
(Tertiary)

Halogenoalkanes are used to prepare alcohols, ethers, alkenes, alkynes and amines by substituting the halogen with appropriate groups or elimination of hydrogen halide. They are also used as solvents, refrigeration gases and insecticides.

3.5 Preparing halogenoalkanes

Halogenoalkanes can be prepared starting from alkanes, by direct substitution of hydrogen by chlorine or bromine, by addition of a halogen or hydrogen halide to alkenes or alkynes, and by treating alcohols with halogenating agents such as hydrogen halides, phosphorus halides and thionyl chloride.

1. Halogenoalkanes from alkanes

In the previous section, we have already discussed the chlorination of methane and the reaction of bromine with propane. The direct halogenation of alkanes is not often used as a preparative method for alkyl halides, since mixtures of halogenated products are formed and it is difficult to control the reaction to give a specific product.

2. From alkenes and alkynes

Hydrogen halides react with alkenes and alkynes to give halogenoalkanes. For example, hydrogen bromide reacts with ethene to give bromoethane.

$$\underset{\text{Ethene}}{CH_2{=}CH_2} + HBr \longrightarrow \underset{\text{Bromoethane}}{CH_3CH_2Br}$$

3. From alcohols

Alcohols react with hydrogen halides, PCl_3, PCl_5, PBr_3, PI_3 and $SOCl_2$ to give corresponding halogenoalkanes.

$$\underset{\text{1-Propanol}}{CH_3CH_2CH_2OH} + HBr \longrightarrow \underset{\text{1-Bromopropane}}{CH_3CH_2CH_2Br} + H_2O$$

$$\underset{\text{Ethanol}}{3CH_3CH_2OH} + PCl_3 \longrightarrow \underset{\text{Chloroethane}}{3CH_3CH_2Cl} + H_3PO_3$$

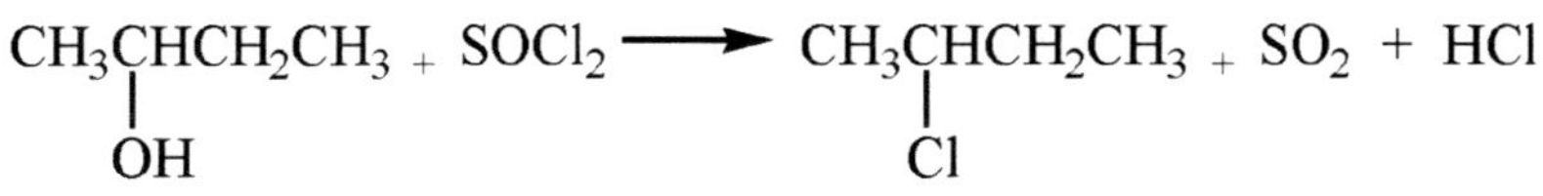

2-Butanol 2-Chlorobutane

3.6 The reactions of halogenoalkanes

Halogenoalkanes are generally polar compounds. They undergo two important types of reactions: nucleophilic substitution and β-elimination.

Nucleophilic substitution

This is a reaction in which the halogen atom in a halogenoalkane is substituted by a nucleophile. A nucleophile is a molecule or an anion which has a pair of electrons that can bond with another atom. A nucleophile can be represented by Nu: or Nu:$^-$ Examples of nucleophiles are:

$$:\overset{\bullet\bullet}{\underset{\bullet\bullet}{O}}{}^{-}\!-H,\quad :\overset{\bullet\bullet}{\underset{\bullet\bullet}{O}}{}^{-}\!-R,\quad :\overset{\bullet\bullet}{\underset{\bullet\bullet}{Br}}{}^{-}:,\quad H_2\overset{\bullet\bullet}{O}:,\quad :NH_3\ \text{and}\ R\text{-}\overset{\bullet\bullet}{\underset{\bullet\bullet}{O}}\text{-}H$$

The reaction between 2-bromopropane and a hydroxide ion is an example of nucleophilic substitution. When 2-bromopropane is treated with sodium hydroxide in a suitable solvent like aqueous ethanol, 2-propanol is formed. In this reaction, bromine is substituted by the nucleophile, hydroxide as follows:

$$\underset{\text{2-Bromopropane}}{CH_3\underset{\substack{|\\ Br}}{C}HCH_3} + OH^- \longrightarrow \underset{\text{2-Propanol}}{CH_3\underset{\substack{|\\ OH}}{C}HCH_3} + Br^-$$

These reactions are important in the preparation of alcohols, ethers and various other compounds as shown in the following reactions. The mechanism of nucleophilic substitution reactions is explained in Section 4.8.

$$\underset{\text{Bromoethane}}{CH_3CH_2Br} + \underset{\text{Sodium methoxide}}{CH_3O^-Na^+} \longrightarrow \underset{\text{Ethylmethylether}}{CH_3CH_2OCH_3} + Na^+Br^-$$

$$\underset{\text{1-Bromopropane}}{CH_3CH_2CH_2Br} + \underset{\text{Sodium cyanide}}{Na^+CN^-} \longrightarrow \underset{\text{Butanenitrile}}{CH_3CH_2CH_2CN} + Na^+Br^-$$

Elimination

When a halogenoalkane containing two or more carbon atoms in the molecule reacts with a base, the halogen atom and a hydrogen atom from the next carbon (β carbon) are removed to form an alkene. This reaction is called **elimination**. It is also called **dehydrohalogenation** since a molecule of hydrogen halide is eliminated. Thus bromoethane eliminates a molecule of hydrogen bromide to form ethene, when treated with potassium hydroxide in ethanol.

$$\underset{\text{Bromoethane}}{CH_3CH_2Br} \xrightarrow[\text{-HBr}]{\text{KOH/ethanol}} \underset{\text{Ethene}}{CH_2{=}CH_2}$$

The mechanism of elimination reaction is explained in Section 4.11.

3.7 Conformations of alkanes and cycloalkanes

In ethane, the carbon atoms can rotate about the carbon-carbon bond since the sigma bond between the carbon atoms is cylindrically symmetrical. As a result, we get different three-dimensional arrangements of the atoms. The different arrangements of atoms formed by the rotation of carbon atoms around a single bond are called **conformation**.

When the carbon-hydrogen bonds are as far away from each other as possible, the arrangement is called the fully **staggered conformation** and when the bonds are as close as possible, the fully **eclipsed conformation**. The staggered structure is more stable (less potential energy) as there is less electron-electron repulsion between C-H bonds, than the eclipsed structure. The difference in potential energy between the two structures is about 12 kJ mol^{-1}. The conformational isomers are represented in two ways, **Sawhorse representation** and **Newman projection**. These representations of the ethane molecule are shown in Figure 3.1.

In the staggered conformation, since the hydrogen atoms and the C-H bonds are as far away as possible, there is least interaction strain present in the molecule from them. We can say, there is least '**steric strain**'. In an eclipsed conformation, the interaction between the hydrogens as well as C-H bonds, produces steric strain.

The carbon atoms of cyclohexane can be arranged in a number of ways and the most stable arrangement is the **chair conformation**. In this arrangement, all C-C-C bond angles are approximately 109.5° and all the hydrogen atoms are in the staggered position. This produces the least angle strain and steric strain.

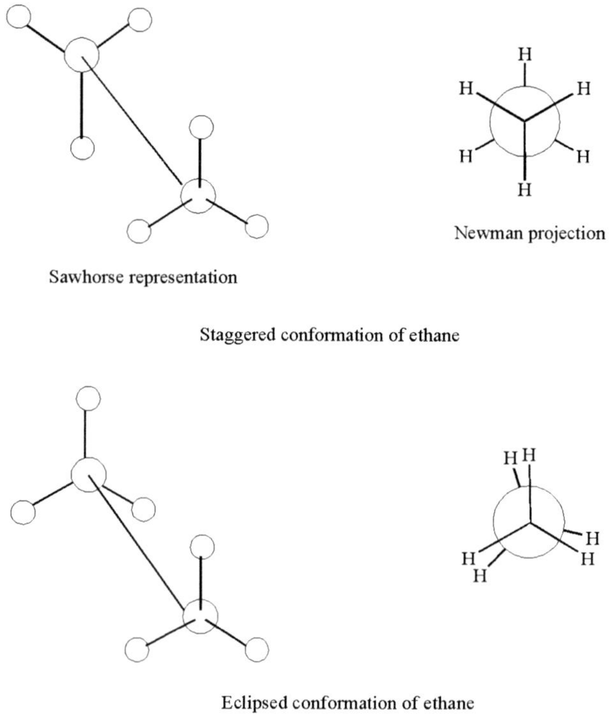

Figure 3.1. Conformations of ethane.

► *Note* – Of the 12 hydrogen atoms in cyclohexane, six are **axial hydrogens** which are perpendicular to the ring or parallel to the axis of the ring, and six are **equatorial hydrogens** which are perpendicular to the axial hydrogens. Note also that three axial hydrogens project upwards and the other three downwards, on alternate carbons.

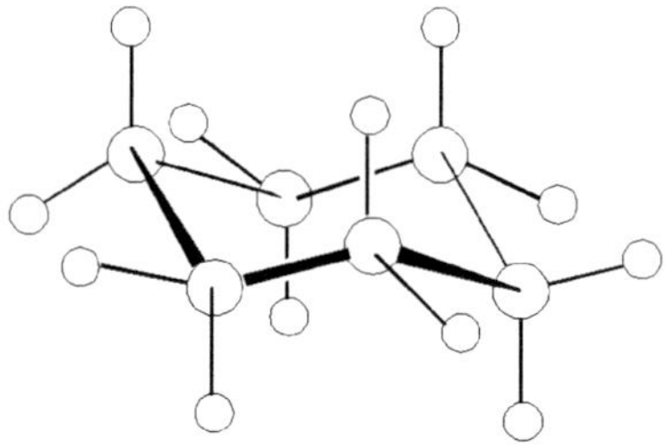

Figure 3.2. Chair conformation of cyclohexane.

One of the other conformations of cyclohexane is the **boat conformation**. This is the least stable conformation because of steric strain. The energy difference between the chair and boat conformations is about 30 kJ mol^{-1}.

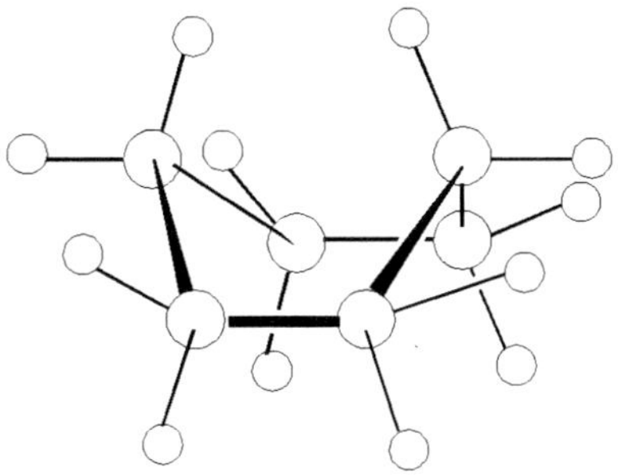

Figure 3.3. Boat conformation of cyclohexane.

► *Hints & tips* – By now, you will have realised the importance of 3-dimensional chemistry. If you can imagine molecules as 3-dimensional objects, you can arrive at a better understanding of the properties and reactions of organic compounds. Models of molecules can be made using model construction kits. It might be fun to get hold of a model kit and construct molecules of methane, ethane, propane, butane, 2-methylpropane, staggered and eclipsed conformations of ethane and chair and boat conformations of cyclohexane.

Tutorial: helping you learn

Progress questions

1. (a) Write an equation for the substitution of a hydrogen atom of ethane by chlorine. Write the IUPAC name of the product.
   (b) Draw the structural formulae and give the IUPAC names of the two possible products when a second hydrogen atom of the reaction product in (a) is replaced by another chlorine atom.

2. Group the following conversions under the reaction types *addition*, *elimination* and *substitution*. Write out the required conditions and equations for the reactions.

(a) 2-Iodopropane to propene
(b) 2-Iodopropane to 2-propanol
(c) 1-Pentanol to 1-bromopentane
(d) 3-Hexene to 3-chlorohexane
(e) 2-Methylpropane to 2-bromo-2-methylpropane
(f) Monobromocyclohexane to cyclohexene.

3. If a mixture of 2,2,3-trimethylpentane and bromine is heated, one major product is formed. Write:

   (a) an equation for the reaction
   (b) the IUPAC name of the product
   (c) the mechanism of the reaction.

   Give your reasoning for the formation of the preferred product mentioned in your answer.

4. Draw the structural formulae of the following compounds:

   (a) 2,3-Dichlorobutane
   (b) 2,3-Dichloro-2,3-dimethylbutane
   (c) 2,2,3,3-Tetrachlorobutane
   (d) 2-Cyclopentylpropane

5. Consider the reaction between ethylbenzene and bromine. Give the formulae of the two monobromo-substituted products formed from side-chain substitution. Which of the two, do you think is the preferred product? Give your reasons.

Discussion point

In the most stable conformation of propane, the bonds are completely staggered. Give representations of this conformation.

Practical assignments

1. Natural gas and petroleum are two important sources of many hydrocarbons. Investigate the composition of natural gas, the fractions of petroleum, and their uses.

2. Halogenoalkanes have a variety of applications. Investigate and give an account of industrially useful halogenoalkanes.

4

Types of Organic Reactions

One-minute summary – Organic reactions can be classified into four categories: addition, substitution, elimination and rearrangement. In this chapter, these reaction types are discussed using appropriate examples. The reaction mechanisms are also explained. Organic molecules are mostly covalent in nature. Factors affecting the breaking and making of covalent bonds, the nature of reagents and reaction conditions are considered here. In this chapter we cover:

- the terms: nucleophile, electrophile, free radical, carbocation and carbanion
- rates of reactions
- nucleophilic and electrophilic addition reactions
- nucleophilic and electrophilic substitution reactions
- elimination reactions
- rearrangement reactions.

4.1 Some useful terms

A nucleophile (nucleus- or positive-liking) is a molecule or an anion which has a lone pair of electrons that can form a bond with another atom. H_2O, ROH, RSH, RNH_2, OH^-, SH^- and RO^- (R stands for alkyl group) are some examples of nucleophiles. In general, *Nu*: or *Nu*:$^-$ is used to represent a nucleophile (See Section 3.6 for the structures of some nucleophiles). Nucleophiles are Lewis bases since they use electron pairs to bond with other substances.

An electrophile (electron-liking) is a molecule or a positive ion which can accept a pair of electrons from a donor and form a bond. An electrophile is represented by E or E^+. Molecules like BH_3 and $AlCl_3$ and ions like H^+ and Br^+ are examples of electrophiles. Electrophiles are Lewis acids since they are electron pair acceptors.

A **free radical** is an atom (or a group of atoms which has an atom) with a single electron. The notation $R^{\cdot}$ can be used to represent a free radical. Free radicals are formed by the **homolysis** (equal cleavage) of a covalent bond. $Cl^{\cdot}$ and $H_3C^{\cdot}$ are examples of free radicals.

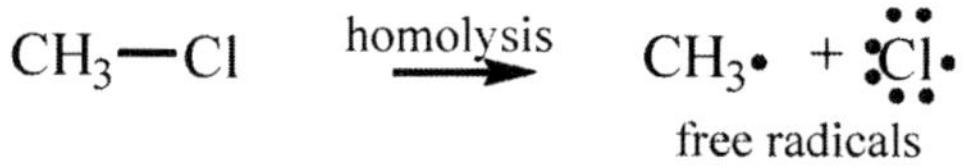

A **carbocation** is an ion which has a positive charge on a carbon atom. A carbocation is formed by the **heterolysis** (unequal cleavage) of a covalent bond.

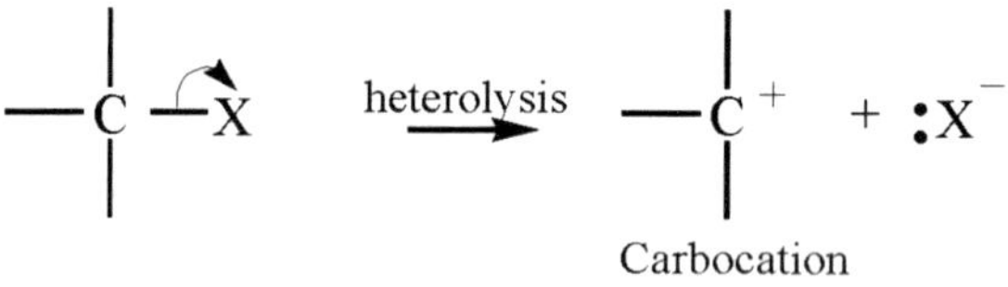

A **carbanion** is an ion which has a negative charge on a carbon atom. A carbanion is also formed by the heterolysis of a covalent bond.

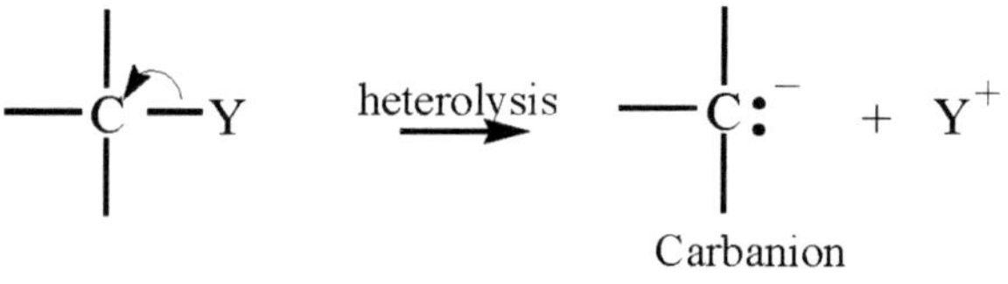

- **A Lewis acid** – is a substance that can accept a pair of electrons, and a **Lewis base** is a substance that can donate a pair of electrons, during bond formation. Lewis acids behave as electrophiles and Lewis bases behave as nucleophiles. But the terms electrophile and nucleophile are used in connection with bonds to carbon.

4.2 Explaining reaction rates

The rate of a reaction tells us how fast or how slowly the reaction takes place. The rate of a reaction is defined as the amount of a substance reacted, or a product formed in unit time. For a reaction between A and B to form the product C,

$$A + B \longrightarrow C$$

the rate of the reaction, R, is given by the rate expression,

$$R = k\,[A]\,[B]$$

where k is called the rate constant. It is a constant for a reaction at constant temperature. The symbol [ ] represents the concentrations of the reactants in $mol.dm^{-3}$.

For every reaction, the rate expression is obtained experimentally and not by looking at the overall stoichiometric equation. If the above reaction takes place in one step as given above, the reaction rate depends on the concentrations of A and B present in the reaction mixture.

But all reactions need not proceed in one step. If a reaction proceeds through a number of steps, the overall reaction rate depends on the slowest step, in other words, the slowest step is the rate determining step. For example, if the above reaction takes place through two steps,

$$A \longrightarrow \underset{\text{intermediate}}{X} \qquad \text{Step 1 (slow)}$$

$$\underset{\text{intermediate}}{X} + B \longrightarrow C \qquad \text{Step 2 (fast)}$$

For the overall reaction: $A + B \longrightarrow C$

The rate of reaction R is: $R = k\,[A]$

4.3 Studying reaction types

Though there are seemingly a large number of organic reactions, most of them belong to a few reaction types. We will now discuss the four main types of reactions: addition reactions, substitution reactions, elimination reactions and rearrangements, in some detail.

4.4 Addition reactions

An **addition reaction** takes place when a substance 'adds on' to another substance to form a compound. Usually, unsaturated compounds containing double or triple bonds undergo addition reactions. For example, when ethene adds a diatomic molecule A-B, the following addition product is formed:

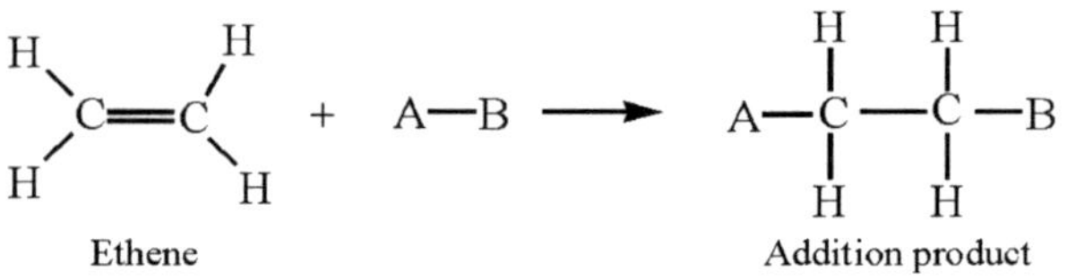

If the substance added is H_2, the product is ethane, C_2H_6. The addition of hydrogen to alkenes or other unsaturated compounds is called **hydrogenation**. A reaction involving the addition of an electrophile is called **electrophilic addition reaction**, and the addition of a nucleophile, **nucleophilic addition reaction**.

4.5 Electrophilic addition reaction

When ethene reacts with hydrogen bromide, bromoethane is formed. This is an example of an electrophilic addition reaction:

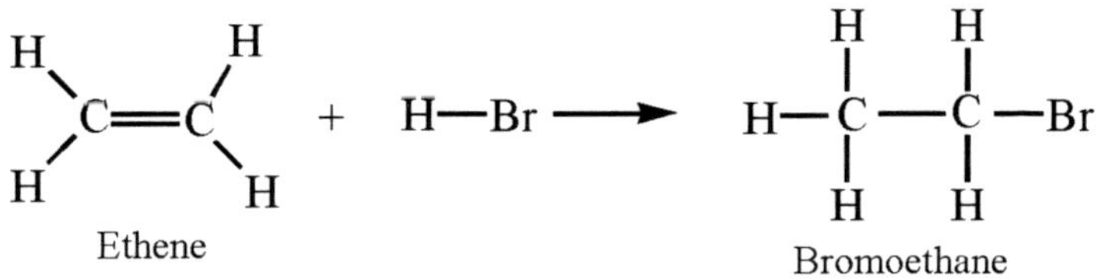

The mechanism of the reaction is as follows: the double bond in ethene consists of a σ bond and a π bond. The π electrons are in lobes above and below the plane of the σ bond. Electrophiles are attracted by the π electrons. Hydrogen bromide is a polar covalent molecule with a δ+ve hydrogen and a δ-ve bromine. The δ+ve hydrogen is attracted by the π electron cloud of the double bond and forms a bond with one carbon using the π electrons. This results in the formation of a carbocation. At the same time the H-Br bond breaks heterolytically to form a bromide ion (step 1).

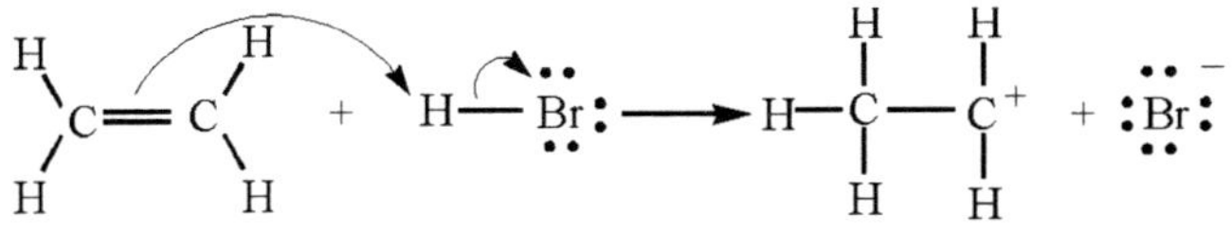

Ethene carbocation (step 1)

▶ *Note* – The movements of the electrons are shown by curved arrows.

In the second step, the reactive carbocation undergoes nucleophilic attack by bromide ion to form bromoethane.

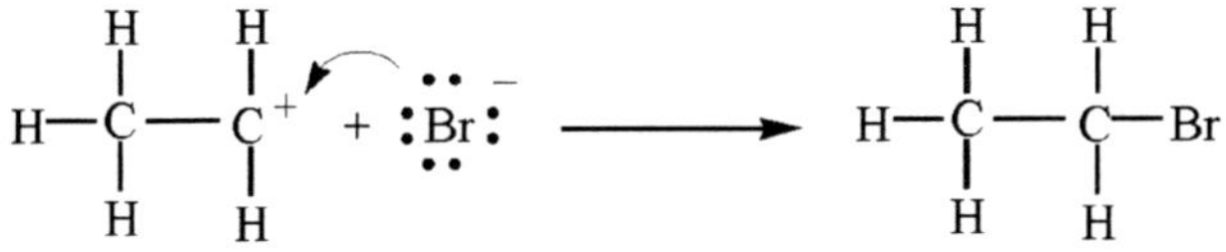

carbocation Bromoethane (step 2)

Let us study the reaction between propene and hydrogen chloride. Here again, an electrophilic addition reaction takes place. In this reaction, there can be two possible products, depending on whether the hydrogen atom is added onto C-1 or C-2 of propene, as given by the following equations:

$$CH_3\,CH{=}CH_2 + \text{H-Cl} \longrightarrow CH_3\,CH_2\,CH_2Cl$$

Propene 1-Chloropropane

$$CH_3\,CH{=}CH_2 + \text{H-Cl} \longrightarrow CH_3\,CHCl\,CH_3$$

Propene 2-Chloropropane

But 2-chloropropane is formed in preference to 1-chloropropane. The product formed in such a reaction can be predicted using **Markovnikov's rule** which states that when a hydrogen halide reacts with an alkene, the hydrogen goes to the carbon which is bonded to more hydrogen atoms. Thus in the above reaction, a hydrogen is added to the carbon of the CH_2- group and not the CH- group. Though Markovnikov's rule helps to predict the product formed in an addition reaction, it does not give the reason for the preferential addition product.

How do we account for this observation? As explained earlier, these reactions take place in two steps. In the first step, a hydrogen ion is added to form a cation. If the hydrogen adds onto C-1 of propene, a secondary cation (1) is formed, and if it adds onto C-2, a primary cation (2) as given below:

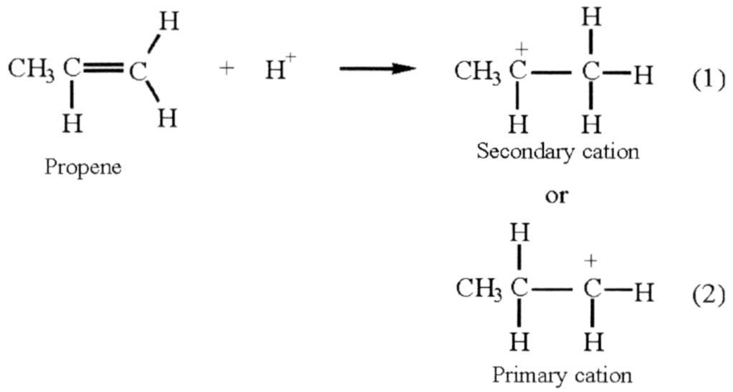

The secondary cation (1) is more stable than the primary cation (2). So cation (1) is formed in preference to cation (2) and it reacts with the chloride ion to give 2-chloropropane as the product.

Notes

1. A primary carbocation is an ion which has a positive charge on a carbon which is bonded to two hydrogen atoms and an alkyl group. A secondary carbocation has one hydrogen and two alkyl groups bonded to the positive carbon and a tertiary carbocation has three alkyl groups bonded to the positive carbon.

2. A tertiary carbocation is more stable than a secondary carbocation, which is more stable than a primary carbocation. The stability decreases in the order

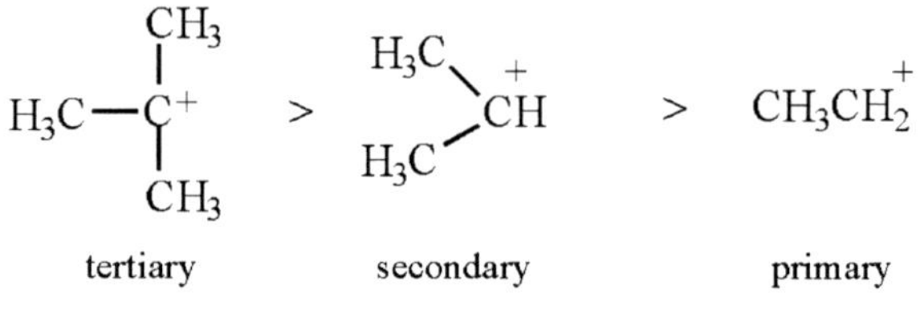

▶ *Remember* – The main reason for the stability of a tertiary carbocation is that the electron releasing inductive effect of the three alkyl groups attached to the positive carbon delocalises or spreads the positive charge. A secondary carbocation spreads the charge less effectively than a tertiary carbocation, but more than a primary one.

4.6 Nucleophilic addition reaction

In this addition reaction, the first step is a nucleophilic attack. Aldehydes and ketones which contain a carbon-oxygen double bond undergo nucleophilic addition reactions with water, alcohols, hydrogen cyanide, ammonia, amines etc. A typical example is the reaction between propanone and hydrogen cyanide.

When propanone is treated with hydrogen cyanide in the presence of sodium cyanide which acts as a catalyst, 2-hydroxy-2-methylpropanenitrile is formed.

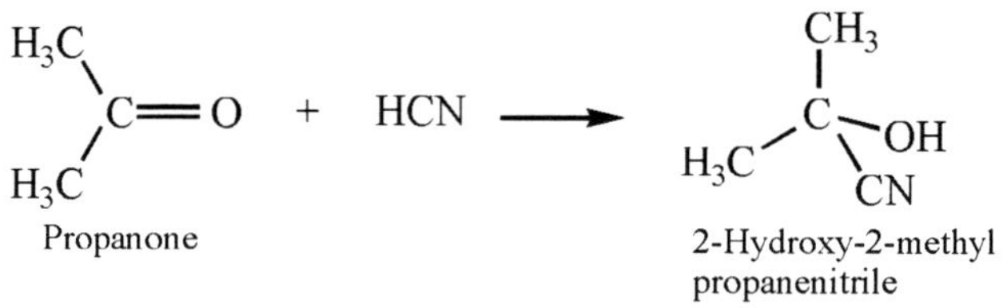

The carbonyl group is polar having a δ+ve carbon and a δ-ve oxygen. The nucleophilic cyanide ion is attracted by the positive carbon of the carbonyl group and forms a bond with it using a pair of electrons of the cyanide carbon. At the same time, a pair of electrons from the carbonyl double bond shifts to oxygen to form the intermediate ion.

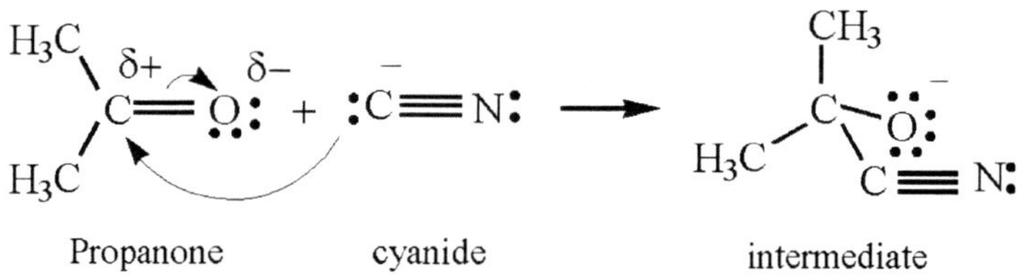

In the second step, the intermediate ion takes a proton from a hydrogen cyanide molecule to form the product and a cyanide ion. In this reaction, in total, a molecule of hydrogen cyanide is added. The function of sodium cyanide is to provide cyanide ions to start the reaction, but these are regenerated in the second step.

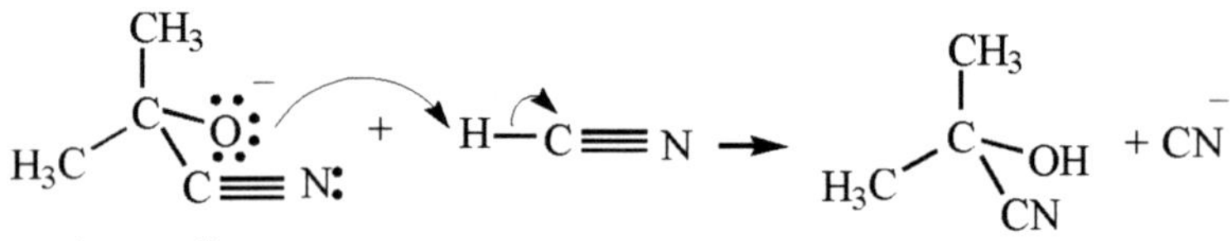

4.7 Substitution reaction

A **substitution reaction** is a reaction in which an atom or a group of atoms in a compound is substituted by another. The halogenation of alkanes and the conversion of halogenoalkanes to alcohols are examples of substitution reactions.

$$CH_3CH_3 + Br\text{-}Br \longrightarrow CH_3CH_2Br + HBr$$

Ethane Bromoethane

$$CH_3CH_2Br + OH^- \longrightarrow CH_3CH_2OH + Br^-$$

Bromoethane Ethanol

Substitution reactions can be classified under three categories, namely **nucleophilic substitution, electrophilic substitution** and **free radical substitution**. The mechanism of free radical substitution has been discussed in Section 3.3.

4.8 Nucleophilic substitution

In this type of reaction, one nucleophile displaces a weaker one. In the conversion of bromoethane to ethanol, the nucleophile OH^- displaces Br^-.

$$CH_3CH_2Br + OH^- \longrightarrow CH_3CH_2OH + Br^-$$

Bromoethane nucleophile Ethanol nucleophile

From the rate determination studies of nucleophilic substitution reactions, it can be seen that there are two types of reaction mechanism. These are the S_N1 and S_N2 mechanisms.

S_N1 mechanism

In S_N1, S stands for *S*ubstitution, N for *N*ucleophilic and 1 for *uni*-molecular. The reaction between 2-bromo-2-methylpropane and a hydroxide ion proceeds by the S_N1 mechanism. The reaction takes place in two steps. In the first step, a bromide ion breaks off from the halogenoalkane to form a carbocation. This step is slow as it requires the breaking of a covalent bond and it is the rate determining step. Since only one molecule is involved in the rate determining step, it is unimolecular.

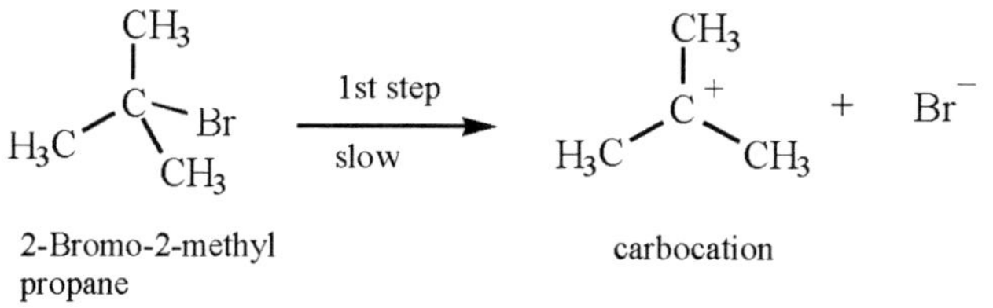

In the second step the carbocation combines with the nucleophile, in this case the hydroxide ion, to form the product, 2-methylpropan-2-ol.

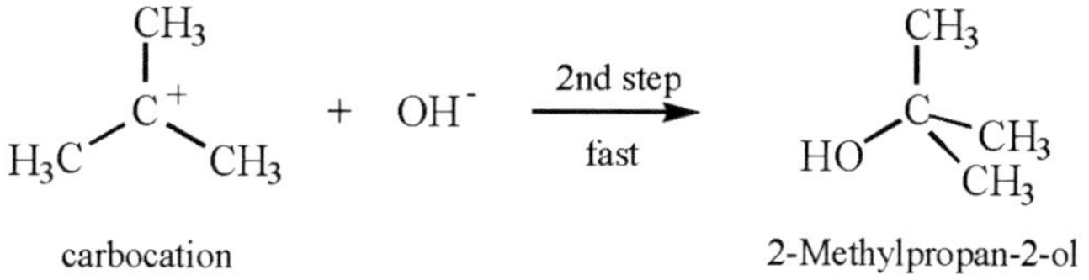

S_N2 mechanism

In S_N2, S stands for *S*ubstitution, N for *N*ucleophilic and 2 for *bi*molecular. The substitution reaction between bromoethane and a hydroxide ion to form ethanol takes place by the S_N2 mechanism. In this reaction, the nucleophilic hydroxide ion is attracted by the δ+ve carbon of the halogenoalkane and forms a weak bond with it, resulting in the formation of an intermediate ion. In the intermediate ion, both hydroxide and bromide are bonded to the carbon by weak bonds.

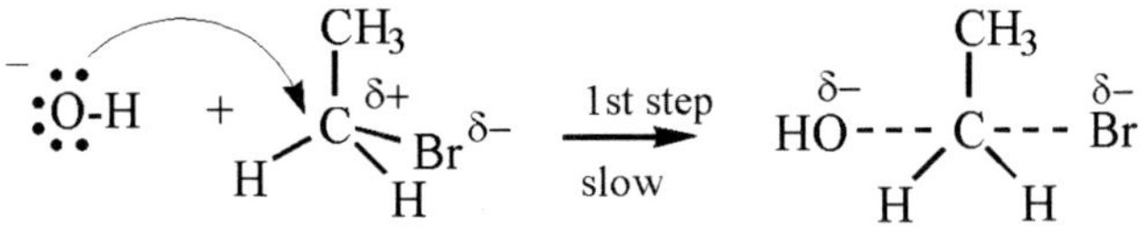

This step is slow, and so is the rate determining step. Since two reactant species (alkyl halide and OH^-) are involved in the rate determining step, it is a bimolecular reaction and the mechanism is S_N2. The intermediate loses a bromide ion in the second step of the reaction to give the product.

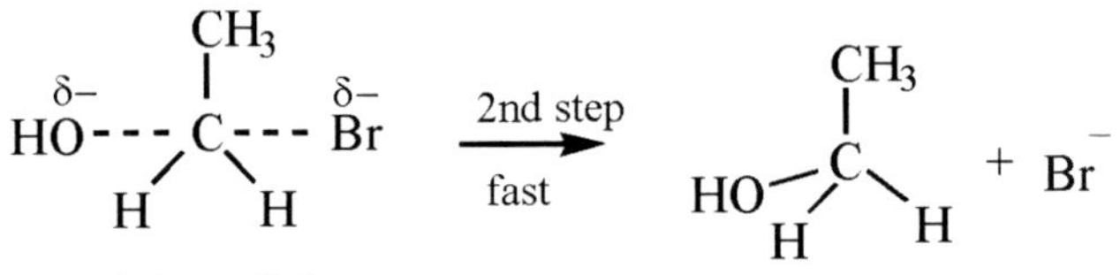

4.9 Factors affecting the rates of S_N1 and S_N2 reactions

We have seen above that one alkyl halide reacts with OH^- by the S_N1 mechanism and another alkyl halide by the S_N2 mechanism.

The S_N1 mechanism requires the breaking of the carbon-halogen bond to form a carbocation and a halide ion. Formation of a stable carbocation favours this step of the reaction. As explained earlier, since a tertiary carbocation is more stable than secondary and primary carbocations, tertiary alkyl halides react by the S_N1 mechanism.

The use of a **protic** solvent (a solvent like water or ethanol, which has a hydrogen atom bonded to a very electronegative atom) helps the dissociation of the alkyl halide, as the cation and anion formed can solvate.

In an S_N2 mechanism, an alkyl halide and a nucleophile are involved in the rate-determining step. The nucleophile attacks the positive carbon attached to the halogen through the side opposite to the halogen. The reaction can be fast if there is no hindrance to this attack. A tertiary alkyl halide with three bulky alkyl groups does not allow the reaction between the carbon and the nucleophile to develop easily. A methyl halide is the least hindered and so the most reactive by the S_N2

mechanism, followed by primary alkyl halides and then secondary halides.

The reactivity of the nucleophile and its concentration influence the S_N2 mechanism. Large anions are more effective nucleophiles in a protic solvent. Small anions are strongly hydrogen-bonded with the solvent and so less effective. In an **aprotic** solvent (a solvent which has no hydrogen atoms attached to the electronegative atom) like dimethyl sulphoxide (DMSO), a small anion is a better nucleophile as the anion is not solvated.

Considering the above facts, it can be seen that tertiary alkyl halides generally undergo S_N1 reactions, and the tendency for such a reaction decreases from tertiary to secondary to primary alkyl halides. On the other hand a methyl and a primary halide usually react by the S_N2 mechanism, a secondary halide by mixed mechanisms.

4.10 Electrophilic substitution

The reaction between chlorine and benzene in the presence of iron (III) chloride or aluminium chloride to form chlorobenzene is an example of electrophilic substitution.

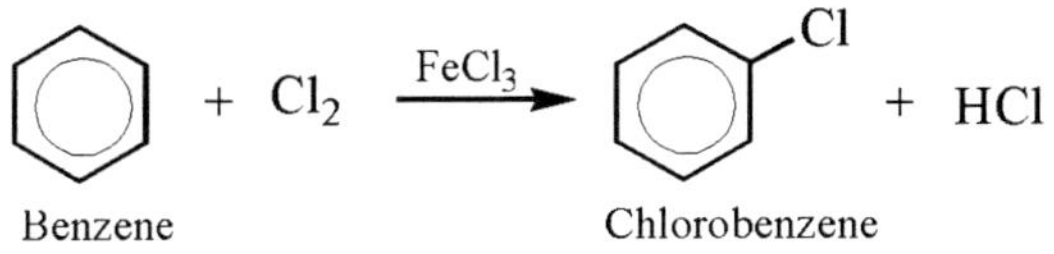

Benzene has three pairs of π electrons which form a negatively charged electron cloud above and below the carbon-carbon ring thus allowing electrophiles to react with benzene. First, the chlorine molecule is polarised in the presence of benzene and $FeCl_3$ so that one end of the molecule becomes positive and the other end negative. The positive end of the chlorine is attracted by the negative electron cloud of benzene and a bond is formed with Cl^+ using a pair of π electrons to form an intermediate carbocation, while at the same time, Cl^- adds onto $FeCl_3$ to form $FeCl_4^-$. This is shown by the following equation.

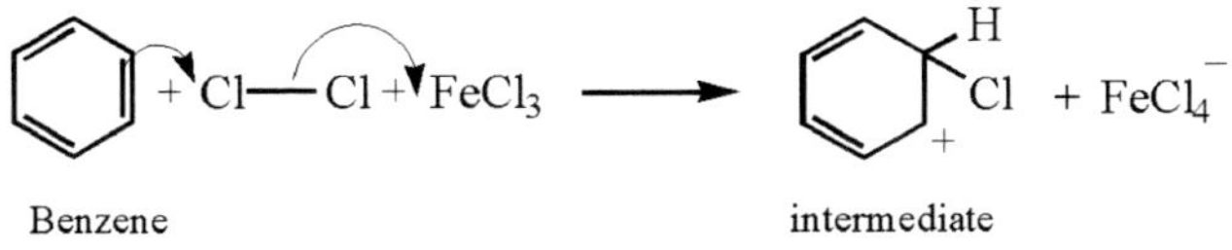

The intermediate carbocation is a resonance hybrid of three contributing forms:

canonical or contributing forms of the intermediate carbocation

A hydrogen ion is lost from the intermediate to form chlorobenzene thus restoring the six π electron cloud.

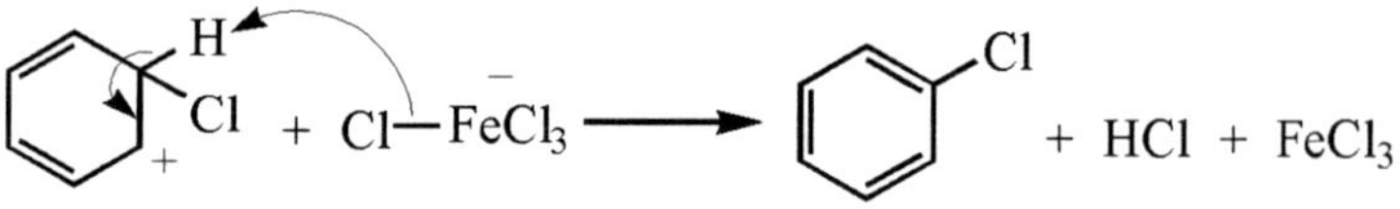

intermediate　　　Chlorobenzene

Even though there are three double bonds or six π electrons in benzene, here benzene undergoes substitution and not addition. This is because the reaction path leading to the restoration of aromaticity, i.e. abstraction of a proton by Cl^- is of lower energy than that leading to addition, i.e. attack of Cl^- on carbon. So this reaction pathway is due to the effect of aromatic stabilization.

4.11 Elimination

Elimination is a reaction in which two atoms or groups of atoms are removed from adjacent carbon atoms of a molecule. The conversion of a halogenoalkane to an alkene and an alcohol to an alkene are examples of elimination reactions. The former reaction is called **dehydrohalogenation** as a molecule of hydrogen halide is eliminated, while the latter is called **dehydration** as a molecule of water is lost.

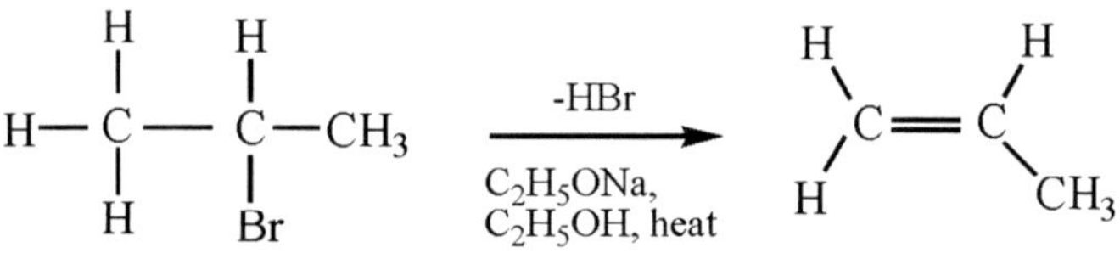

2-Bromopropane　　　Propene

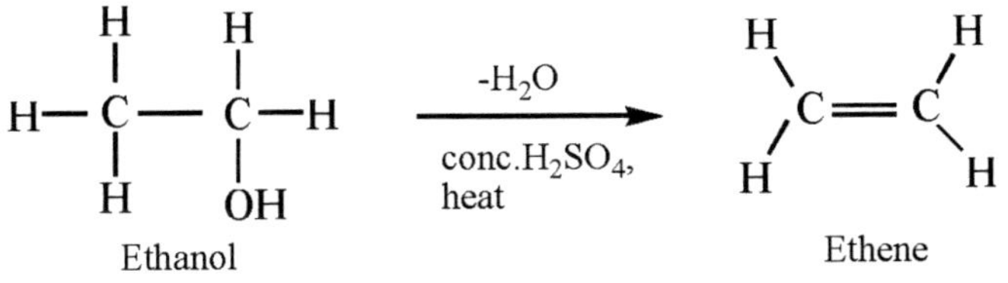

The mechanism of elimination reactions can be either El (*E*limination, *uni*molecular) or E2 (*E*limination, *bi*molecular) as in the case of nucleophilic substitution reactions.

E1 mechanism

In an El reaction, as in an S_N1 reaction, an alkyl halide first dissociates to form a carbocation. When 2-methylpropene is formed from 2-chloro-2-methylpropane in a mixture of ethanol and water, the reaction goes through an El mechanism. In the first step of this two-step reaction, the reactant dissociates into a carbocation and chloride. This step is slow and thus the rate determining step. Since only one molecule of the reactant is involved in this step, it is unimolecular.

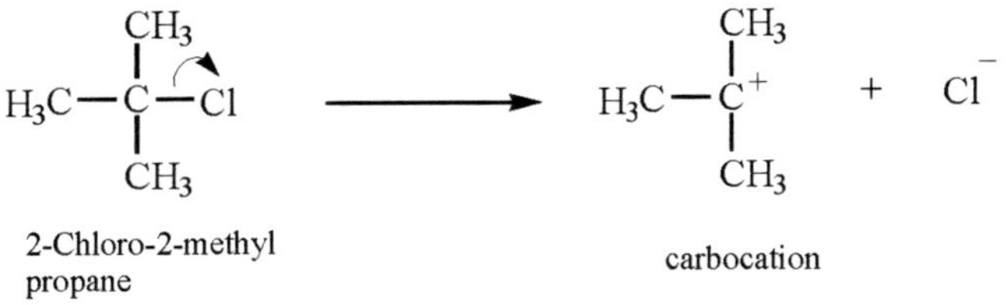

In the second step, the solvent acts as a base, removing a proton from the next carbon so that a pair of electrons shifts to form a double bond.

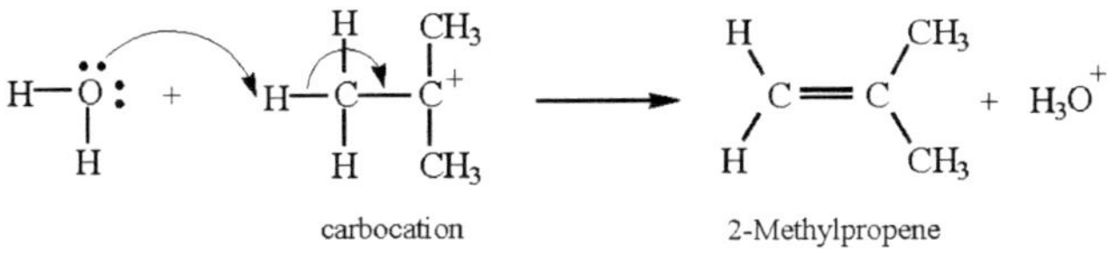

E2 mechanism

Many elimination reactions occur by E2 mechanism. When 2-bromo-propane is treated with sodium ethoxide in ethanol, propene is formed. Experimental studies show that the rate of this reaction depends on the concentrations of 2-bromopropane and ethoxide ion. This is a one-step reaction, involving two reactant species, hence bimolecular. Ethoxide ion acts as a base and starts to bond with a hydrogen atom bonded to the β carbon atom. At the same time, the bond between that hydrogen and the carbon shifts to form a double bond, and a bromide ion breaks off as shown below.

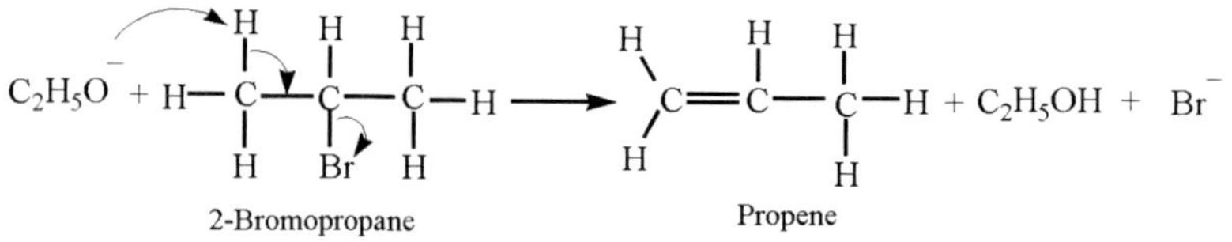

4.12 Nucleophilic substitution versus elimination

1. Methyl halides and primary halides usually undergo S_N2 reactions with nucleophiles or bases. But with a hindered base like *tert*-butoxide, $(CH_3)_3CO^-$, E2 elimination takes place.
2. Secondary halides usually undergo S_N2 reactions with weak bases like Cl^-, Br^-, I^-, CN^- and $CH_3CO_2^-$ and E2 reactions with strong bases like $CH_3CH_2O^-$ and NH_2^- and with hindered bases like $(CH_3)_3CO^-$.
3. Tertiary alkyl halides undergo S_N1 reactions at low temperatures and E1 reactions at high temperatures. E2 elimination predominates with a strong base and high temperature.

4.13 Rearrangement reactions

Some chemical reactions yield products that are unexpected. A typical example is the reaction between hydrogen chloride and 3-methyl-1-butene. The expected product is 2-chloro-3-methylbutane. But the reaction gives a mixture of 40% of the expected product and 60% of 2-chloro-2-methylbutane.

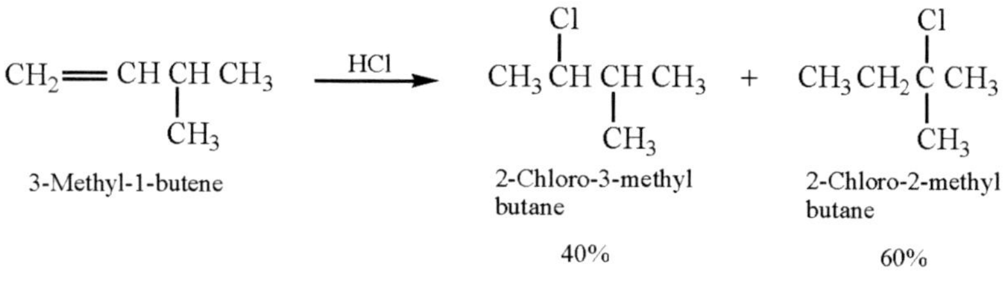

In this reaction, rearrangements take place. This is because the intermediate cation formed in the first step of the reaction stabilises by rearranging as shown below.

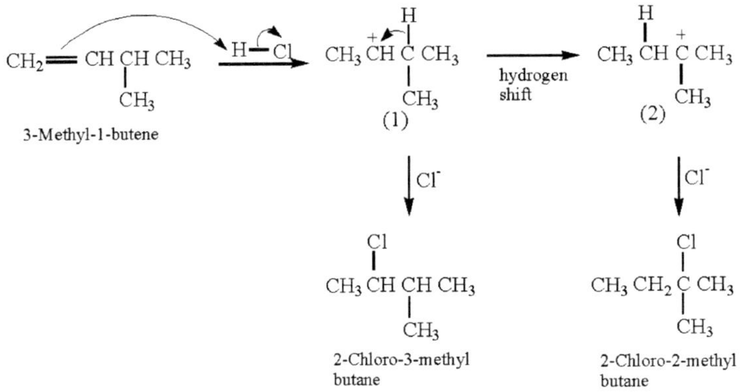

C-1 of the reactant bonds to the hydrogen of HCl using a pair of electrons from the double bond to form the intermediate carbocation (1). Some of the cations (1) add onto chloride ions to form the product 2-chloro-3-methylbutane. A large proportion of the cations (1) rearrange by shifting a hydrogen from C-3 to C-2, to form the more stable tertiary cation (2), which reacts with the chloride ion to give the product 2-chloro-2-methylbutane.

When 3,3-dimethyl-2-butanol is heated with phosphoric acid, dehydration takes place to form a mixture of alkenes. 97% of the product is rearranged alkenes and only 3% is the expected product.

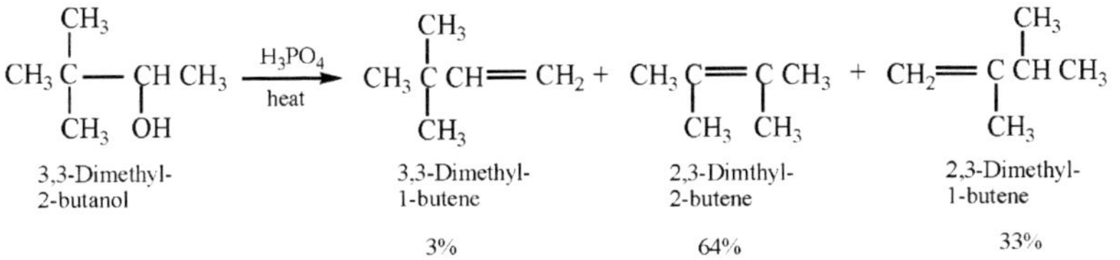

In the first step of this reaction, the reactant takes a proton from the acid catalyst and eliminates a water molecule to form a secondary carbocation. A proton is lost from C-1 of this carbocation to give the product 3,3-dimethyl-1-butene.

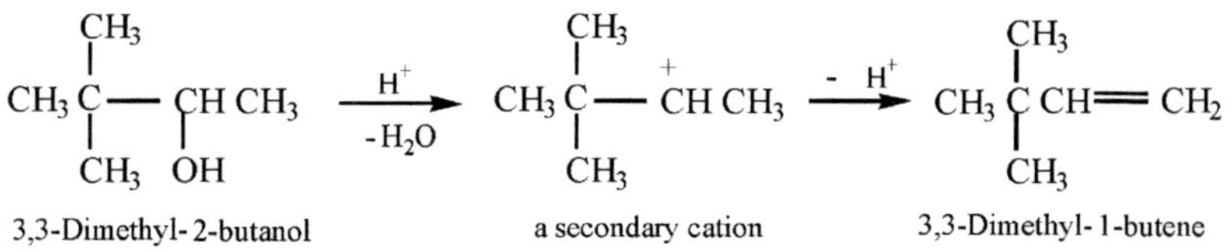

However, most of these secondary cations rearrange immediately by methyl shift to form more stable tertiary cations as shown below.

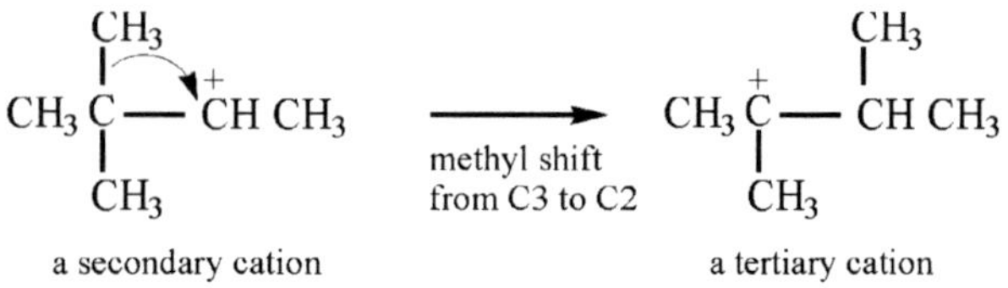

The tertiary cation eliminates a proton from C-1 to form 2,3-dimethyl-1-butene, or from C-3 to form 2,3-dimethyl-2-butene.

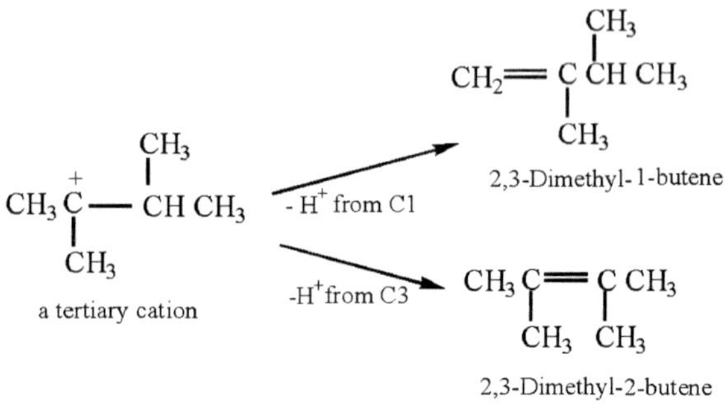

Tutorial: helping you learn

Progress questions

1. For the isomeric carbocations of $C_5H_{11}{}^+$, draw structural formulae of
   (a) three primary carbocations
   (b) two secondary carbocations
   (c) one tertiary carbocation.

2. (a) What is the major reaction product, when 1-bromopentane reacts with each of the following bases?
   (i) hydroxide ion (HO^-)
   (ii) *tert*- butoxide ion ($(CH_3)_3CO^-$)
   (b) Explain the mechanism of reaction in each of the above cases.

3. What products would you expect from the following reactions? Give the reaction conditions and equations.
   (a) Addition of hydrogen chloride (HCl) on 2-methyl-2-butene.
   (b) Addition of hydrogen cyanide (HCN) on propanal (CH_3CH_2CHO).
   (c) Reaction of bromocyclohexane ($C_6H_{11}Br$) with ethoxide ($CH_3CH_2O^-$).
   (d) Reaction of benzene with bromine in the presence of aluminium bromide.

Discussion points

Discuss and give explanations for the following observations:

1. Substitution reactions of 2-bromo-3,3-dimethylbutane are slow, but in the presence of a base, elimination is readily achieved.
2. Reaction of 2-bromo-2-methylpropane with hydroxide ion (HO^-) in methanol at 50°C gives 2-methyl-1-propene as the product.
3. Treating 2-bromo-2-methylpropane with methanol at 25°C gives 2-methoxy-2-methylpropane as the major product and 2-methyl-1-propene as the minor product.

Practical assignments

Conduct an investigative study of the factors which influence reaction rates, and explain these in terms of molecular collisions and energy changes.

Study tips

Reaction mechanism studies help you to understand why and how reactions occur. For every reaction you come across, try to give an explanation of why it takes place and what the mechanism is. Use curved arrows to show the movement of electrons in explaining the reaction mechanism.

5

Isomerism and Stereochemistry

One-minute summary – In this chapter we are going to cover the three-dimensional nature of organic molecules. A single molecular formula of a compound gives the number and type of atoms in one molecule. A molecular formula can stand for a number of compounds called isomers. There are constitutional isomers in which atoms are connected differently within their molecules, and stereoisomers in which molecules differ only in the spatial arrangement of their atoms. Geometrical isomers are stereoisomers arising from restricted rotation about a double bond or a ring. There are also pairs of stereoisomers which are mirror images just as the right hand is a mirror image of the left. These isomers are not superimposable onto one another. In this chapter we shall study:

- constitutional isomers
- chirality and asymmetric carbon atoms
- optical activity
- perspective and Fischer projection formulae
- priority rules for substituents
- the R-S system of naming enantiomers
- enantiomers, diastereomers and meso compounds
- geometrical isomers
- the Z-E system of naming geometrical isomers.

5.1 Isomerism: constitutional isomers

Isomers are compounds with the same molecular formula but yet different. We have seen earlier that there are two alkanes with the same molecular formula C_4H_{10}, but different arrangements of atoms in the molecule. These compounds are examples of constitutional isomers.

- **Constitutional isomers** – are isomers that have different arrangements of atoms in their molecules. Butane and 2-methylpropane are constitutional isomers.

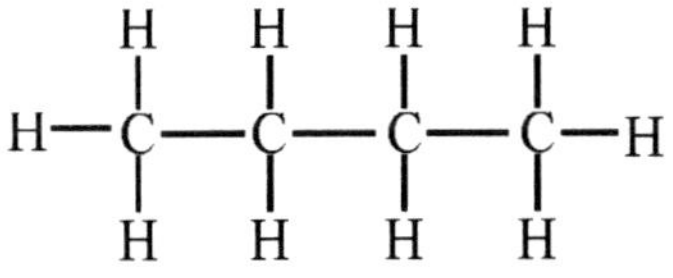

Butane

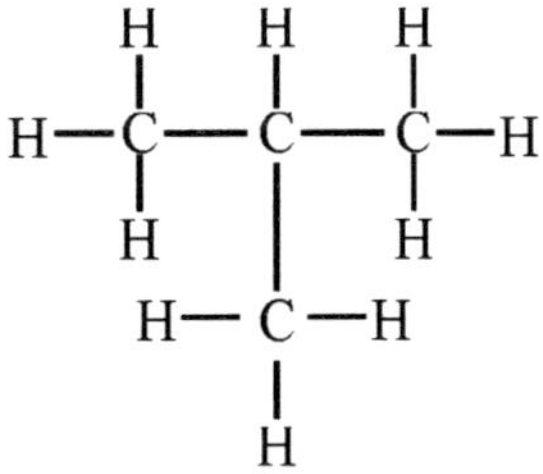

2-Methylpropane

Propanal, C_3H_6O and propanone, C_3H_6O are constitutional isomers.

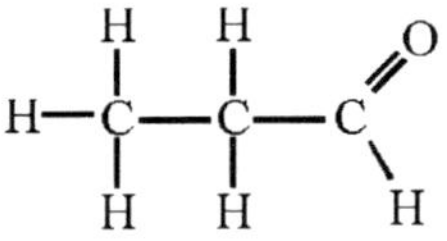

Propanal

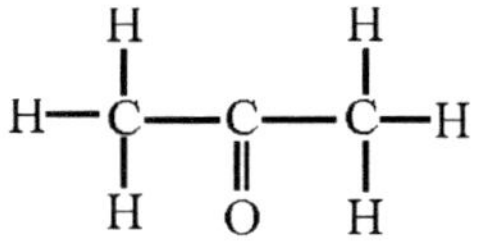

Propanone

5.2 Stereoisomers

Stereoisomers are compounds with the same arrangement of atoms in their molecules but which differ in the arrangement of their atoms *in space*. An example of a compound which can exhibit stereoisomerism is 2-bromobutane. The atoms in 2-bromobutane, $CH_3CHBrC_2H_5$, can be arranged in two different ways:

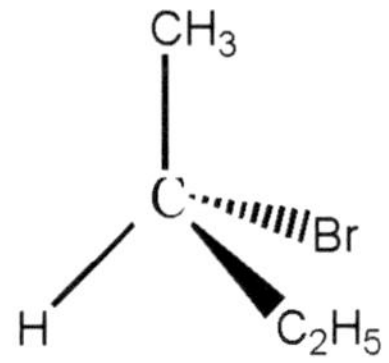

(1)

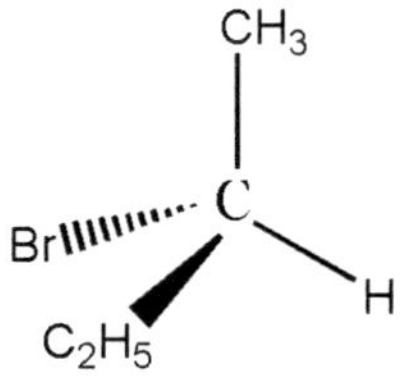

(2)

Figure 5.1. The enantiomers of 2-bromobutane.

Remember that molecules are three-dimensional objects. C2 of 2-bromobutane bonds tetrahedrally with a hydrogen atom, a methyl group, a bromine atom and an ethyl group in the same order in both structures (1) and (2). Structure (1) is not superimposable on structure (2); however, (2) is the mirror image of (1). They are stereoisomers, and stereoisomers which are mirror images of one another are called **enantiomers**.

▶ *Key point* – In the perspective formula of 2-bromobutane (Figure 5.1), H- and CH_3- are on the plane of the paper, C_2H_5- above and Br- beneath.

An enantiomer has a carbon atom which is bonded to four different atoms or groups of atoms. Such a carbon atom is referred to as an **asymmetric carbon**, **asymmetric centre**, **chiral carbon** or **chiral centre**. A molecule which is superimposable on its mirror image is not chiral; it is **achiral**. A chiral carbon atom is sometimes denoted by an *asterisk*.

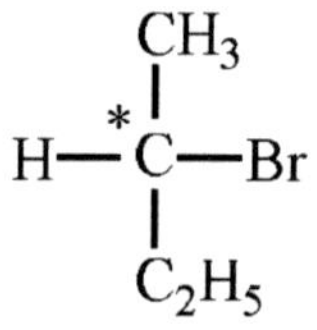

2-Bromobutane

The enantiomers of a compound have the same melting and boiling points, the same solubility in common solvents and the same density. But they differ in *optical activity*.

▶ *Key point* – A chiral molecule has no plane of symmetry, whereas an achiral molecule possesses a plane of symmetry. A plane of symmetry is an imaginary plane dividing the molecule in such a way that one half of the molecule is the reflection of the other half.

5.3 What is optical activity?

If a beam of *plane-polarised light* is passed through a solution of an enantiomer containing a chiral centre, the plane of polarisation rotates. If one enantiomer rotates the plane of polarised light in a clockwise direction, its mirror image isomer rotates it in an anti-clockwise direction. This property of enantiomers to rotate the plane of polarised light is called **optical activity**.

Polarimeter

A polarimeter is a device used to measure the **angle of rotation**, α, of plane-polarised light by an optically active compound. It consists of a light source (usually a sodium lamp), a polariser (a fixed Nicol prism), a tube for holding the optically active substance or its solution, an analyser (another Nicol prism which can be rotated) and a scale to note the angle of rotation. The light from the sodium lamp is passed through the polariser to get plane polarised light which is then passed through the tube. If the tube is empty or if it contains an optically inactive substance, maximum light can be observed and the scale reads 'zero' angle. But if the tube contains an optically active substance, the plane of polarised light rotates and a diminished light is observed. The analyser is now rotated through the required angle to view the maximum light and the angle of rotation can be read from the scale.

The angle of rotation, α, depends on the concentration of the solution, the length of the tube, the temperature and the wavelength of the light. For a light source of fixed wavelength (for example, a sodium lamp) and at a constant temperature, α depends on the concentration and the length of the tube. Chemists use the term *specific rotation,* $[\alpha]$, which is given by

$$[\alpha] = \frac{100\alpha}{l\,c}$$

where: α = observed rotation
c = concentration of the solution of the substance in g/100 cm^3
l = length of the tube in decimetres

If an enantiomer rotates the plane of polarised light in a clockwise direction it is said to be positive (+) or **dextrorotatory**. Its mirror-image enantiomer rotates light in an anti-clockwise direction, thus negative (-) or **levorotatory**. Solutions of enantiomers of the same concentration rotate through equal angles but in opposite directions. A solution which contains equal amounts of two enantiomers does not rotate light, since the effect of the rotation caused by one isomer is cancelled by that of the other. Such a mixture is called a **racemic mixture**.

5.4 The Fischer projection formula

Since 1891 the Fischer projection formula has been used to draw the structures of enantiomers. In this representation, two atoms or groups of atoms attached to an asymmetric carbon atom by horizontal lines lie above the plane of the paper, and the two attached by vertical lines lie below the plane of the paper. The Fischer projection formulae of the enantiomers of 2-butanol are given below:

(-) - 2-Butanol (+) - 2-Butanol

Figure 5.2. The Fischer projection formulae.

5.5 The R-S system of naming enantiomers

It is important to know how substituents are arranged in the order of **priority** before we can name enantiomers by the R-S system. The rules necessary to decide priority are outlined below.

1. First, find the atoms attached to the chiral carbon. The atom with the highest atomic number has the highest priority. Other atoms are arranged in the order of decreasing atomic number. For example, in 1-chloroethanol, $CH_3CHClOH$, C-1 is attached to CH_3-, Cl-, H- and OH-. The order of priority is

$$\underset{(17)}{Cl} > \underset{(8)}{OH} > \underset{(6)}{CH_3} > \underset{(1)}{H}$$

2. If two of the vicinal atoms (atoms attached to the chiral carbon) are isotopes of the same element, the one with the higher mass number has higher priority. Thus ^2H has higher priority than ^1H.

3. If two of the vicinal atoms are the same, then atoms attached to the vicinal atoms are examined to decide the order of priority and this

process is continued until the first point of difference is found. For example, in 2-bromobutane, the chiral carbon is bonded to two other carbon atoms as shown below. In CH_3-, the carbon is attached to three hydrogens and in CH_3CH_2-, the vicinal carbon is attached to two hydrogenous and one carbon. So CH_3CH_2- has higher priority than CH_3-.

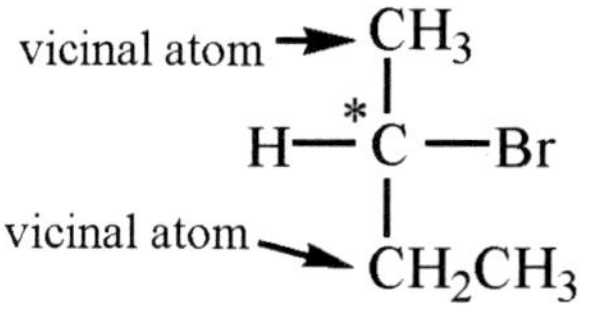

2-Bromobutane

4. If the substituents have atoms attached with double or triple bonds, these atoms are counted as though they are attached doubly or trebly. Note the following examples:

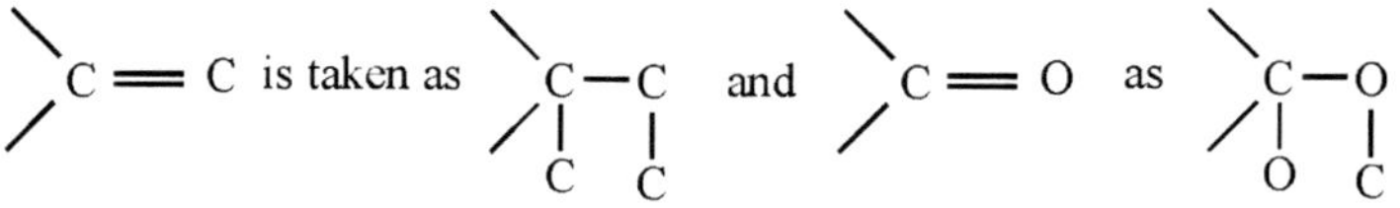

The IUPAC rules for assigning the configuration of chiral compounds by the R-S system are:

(a) Find the chiral centre of the molecule and identify the four substituents attached to the chiral carbon.

(b) Arrange the substituents in the order of decreasing *priority*.

(c) In the tetrahedral model, the atom or group of lowest priority is directed away from you and the other three groups are directed towards you. The isomer in which these three groups in decreasing order of priority are arranged in a clockwise direction is designated as the (*R*)- isomer (from the Latin word *rectus*, right), and the other one, the (*S*)- isomer (from *sinister*, left).

These rules are now applied to find the (R) and (S) isomers of 2-bromobutane. The decreasing order of priority of the substituents attached to the chiral carbon is Br > C_2H_5 > CH_3 > H. These groups are arranged according to the above rules to form the isomers.

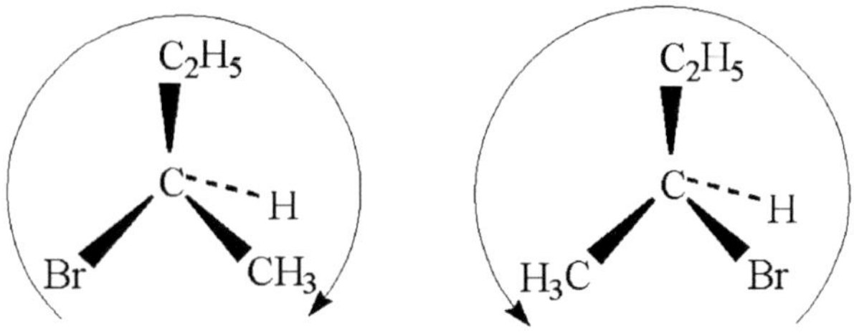

(R)-2-bromobutane (S)-2-bromobutane

Figure 5.3 R-S representations of 2-bromobutane.

5.6 Enantiomers and diastereomers

A molecule containing one chiral centre prepared under normal conditions consists of two isomers called enantiomers. If there are two or more chiral centres in a molecule, the number of possible stereoisomers can be obtained using the relation $\mathbf{2^n}$ where **n** is the number of chiral atoms in the molecule. So if there are two chiral atoms, there can be $2^2 = 4$ stereoisomers.

3-Bromo-2-butanol is an example of a compound with two chiral carbon atoms.

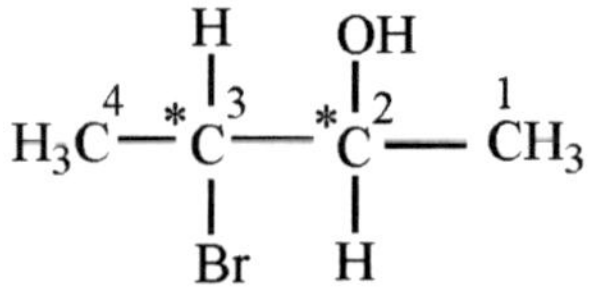

3-Bromo-2-butanol

The stereoisomers of 3-bromo-2-butanol are given below.

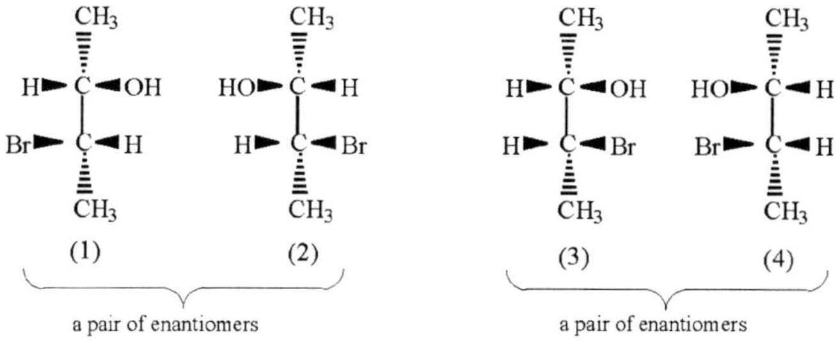

Figure 5.4. Stereoisomers of 3-bromo-2-butanol.

Structures (1) and (2) are mirror images, and they form a pair of enantiomers. Mirror images (3) and (4) form another pair of enantiomers. Structures (1) and (3) or structures (2) and (4) are not mirror images/enantiomers; such stereoisomers are called **diastereomers**. Diastereomers are different compounds with different physical properties like melting and boiling points and solubility.

5.7 Meso compounds

A **meso compound** is a stereoisomer which has two chiral centres, but the molecule is achiral because there is a plane of symmetry. An example is meso tartaric acid. Tartaric acid has three stereoisomers: a pair of enantiomers and a meso isomer, as follows:

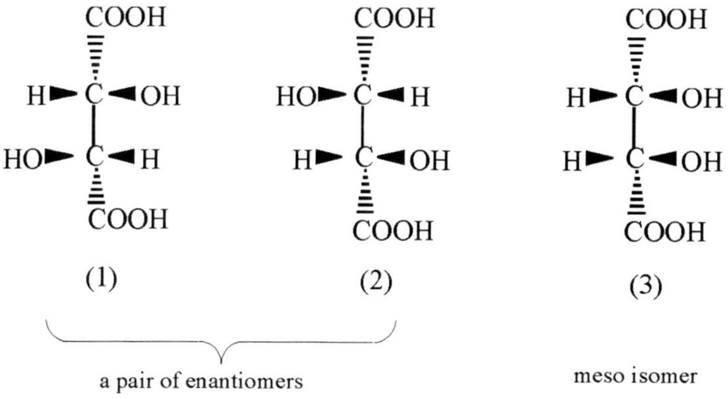

Figure 5.5. Stereoisomers of tartaric acid.

Meso tartaric acid has a plane of symmetry which bisects the molecule through the C2-C3 bond. One half of the molecule is the mirror image of the other half. Thus, if the molecule is rotated through 180^0, its superimposable mirror image is obtained. So meso tartaric acid is achiral. The meso isomer and either of the enantiomers form a pair of diastereomers since they are stereoisomers, but not mirror images.

5.8 Geometrical isomerism

This is another type of stereoisomerism. 2-Butene has two isomers: a *cis*-isomer and a *trans*- isomer. In both isomers, C2 and C3 are bonded to a hydrogen atom and a methyl group. In the *cis*- isomer, the two hydrogen atoms are on the same side of the molecule whereas in the *trans*- isomer, the two hydrogen atoms are on opposite sides. Restricted rotation at the double bond prevents the two compounds from being converted into one another at normal temperatures.

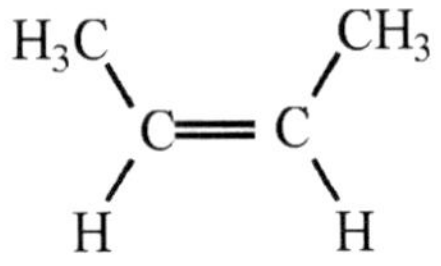

cis-2-butene

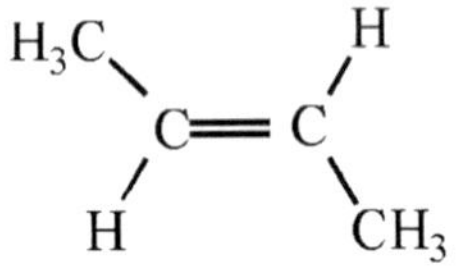

trans-2-butene

This is **cis-trans** or **geometrical isomerism**. The *trans*- isomer is more stable than the *cis*- isomer because there is less strain due to interaction between the bulky alkyl groups when they are spaced apart.

5.9 The Z-E system of naming geometrical isomers

To name geometrical isomers by the Z-E system, decide which of the two groups attached to each carbon of the double bond has higher priority (Section 5.5). If the groups of higher priority are on the same side of the double bond, the alkene is *Z* (from the German word *zusammen*, together). If they are on opposite sides, the isomer is *E* (from *entgegen*, opposite).

In 2-pentene, the priority groups attached to C2 and C3 are a methyl group and an ethyl group respectively. The Z and E isomers of 2-

pentene are:

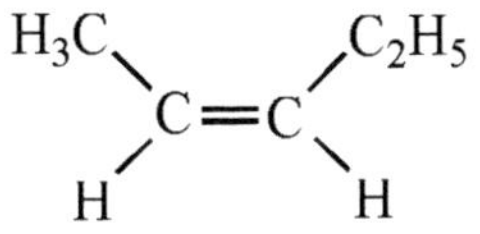

(Z)-2-pentene

(*cis*-2-pentene)

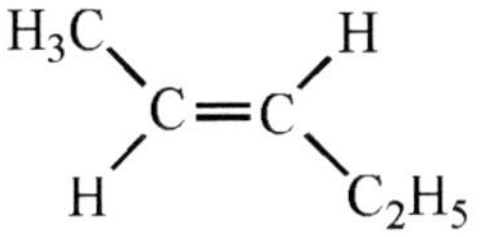

(E)-2-pentene

(*trans*-2-pentene)

Tutorial: helping you learn

Practice questions

For the compound 2-hydroxy-3-hexene:

1. Write the condensed structural formula.
2. Draw the structures of the Z and E isomers.
3. Draw the skeletal formulae of the above isomers.
4. Mark the chiral centre with an asterisk and assign R or S configuration.

Progress questions

1. Arrange the following atoms or groups in order of decreasing priority:

   (a) -Cl, -Br, $-CH_3$, -OH
   (b) -CHO, $-CO_2H$, -H, $-CH_3$
   (c) -H, $-CO_2H$, $-NH_2$, $-C_2H_5$
   (d) -Cl, -H, $-CH_2OH$, $-CH_2CH_3$

2. For the amino acid alanine:

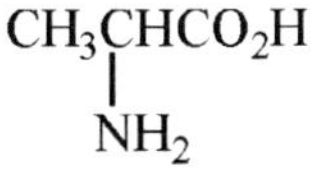

Alanine

   (a) draw a 3-dimensional depiction of the molecule, marking the chiral carbon with an asterisk

(b) draw its mirror-image
(c) give the systematic names, including R and S designations, of the above compounds
(d) draw the Fischer projection formulae of the two compounds.

3. Define the following terms, giving examples:

(a) enantiomers
(b) diastereomers
(c) meso compound
(d) racemic mixture
(e) chiral
(f) achiral

4. (a) Draw the Fischer projections for all the stereoisomers of 2,3-dibromobutane ($CH_3CHBrCHBrCH_3$).
(b) Assign R and S configuration to the chiral centres.
(c) Which of these compounds are optically active and which are not?

5. Assign R or S configurations to the chiral centres, and Z or E to the double bonds in the following molecules:

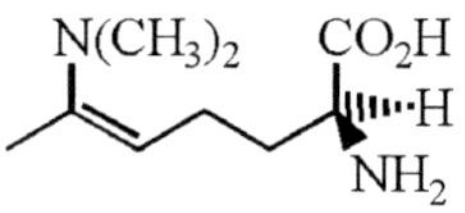

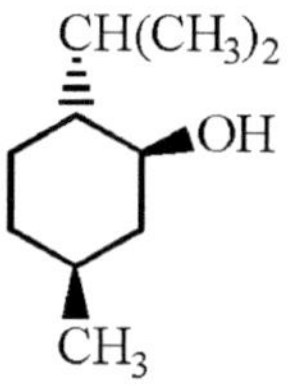

Discussion points
It has been discussed in Section 5.8 that cis-trans isomers exist due to the restricted rotation of the carbon atoms at the double bond. Discuss the possibility of cis-trans isomerism in cycloalkanes, taking 1,2-dimethylcyclopentane as an example.

Practical assignments
Many naturally occurring organic molecules contain one or more chiral carbon atoms. Research for a number of such organic molecules, draw their skeletal structures, mark any chiral centre with an asterisk and predict the number of stereoisomers.

Study tips

Make molecular models of chiral molecules and use the models to assign R and S configurations as explained in Section 5.6. Holding a model in your hand, with two substituents of the chiral carbon facing towards you and two away, practise drawing the Fischer projection formulae.

6

Alkenes, Alkynes and Arenes

One-minute summary – Alkenes and alkynes are unsaturated compounds containing a double bond and a triple bond respectively in their molecules. In both physical and chemical properties, alkenes resemble alkynes closely. They both undergo addition and polymerisation reactions. One main difference between alkenes and alkynes is that terminal alkynes (alkynes which have a triple bond between C-1 and C-2) are acidic, and the hydrogen bonded to the terminal carbon can be substituted. Arenes (aromatic compounds) are also unsaturated and they have a benzene ring as the basic unit in the molecule. The chemistry of the unsaturated compounds is diverse, as they are very reactive and are used in the preparation of industrially important compounds. In this chapter, we shall study:

- methods of preparation of alkenes
- reactions of alkenes
- polymerisation reactions
- preparation of alkynes
- reactions of alkynes
- acidic nature of terminal alkynes
- reactions of benzene and other aromatic compounds.

6.1 Alkenes

Alkenes form a series of compounds with the general formula C_nH_{2n}, where n stands for the number of carbon atoms in the molecule. Alkenes contain one double bond which is equivalent to having two hydrogen atoms less than the corresponding alkane. The first member in the alkene series is ethene, C_2H_4. The bonding in the ethene molecule (Section 1.5), the IUPAC nomenclature of alkenes (Section 2.10), the mechanism of addition reactions of alkenes (Section 4.5) and *cis-trans* isomerism in alkenes (Section 5.8) are discussed in earlier chapters.

Alkenes resemble alkanes in their physical properties. The first few members are gases, and alkenes with five or more carbon atoms are liquids. Alkene molecules are nonpolar, intermolecular forces of attraction thus being van der Waals forces. They are not soluble in water but dissolve in non-polar solvents.

6.2 Methods of preparation of alkenes

1. By the dehydrohalogenation of alkyl halides

Alkenes can be prepared by eliminating a molecule of hydrogen halide from alkyl halides (Section 4.11). When 1-chlorobutane is treated with alcoholic potash (a solution of potassium hydroxide in ethanol), a molecule of HCl is eliminated and 1-butene is formed.

$$\underset{\text{1-Chlorobutane}}{CH_3CH_2\underset{|}{\overset{}{C}}H\underset{|}{C}H_2} \quad (\text{H on C-2, Cl on C-1}) \xrightarrow{\text{KOH, ethanol}} \underset{\text{1-Butene}}{CH_3CH_2CH{=}CH_2}$$

When 2-chlorobutane is treated under similar conditions, a mixture of 1-butene and 2-butene is formed.

$$\underset{\text{2-Chlorobutane}}{CH_3CH_2CHClCH_3} \longrightarrow \underset{\text{1-Butene}}{CH_3CH_2CH{=}CH_2} + \underset{\text{2-Butene}}{CH_3CH{=}CHCH_3}$$

The mechanism of elimination reactions is explained in Section 4.11.

2. By the dehydration of alcohols

Alcohols can be dehydrated in the presence of an acid to form alkenes. Thus, the dehydration of ethanol gives rise to ethene, and 2-butanol to a mixture of butenes.

$$\underset{\text{Ethanol}}{CH_3CH_2OH} \xrightarrow[\text{acid}]{-H_2O} \underset{\text{Ethene}}{CH_2{=}CH_2}$$

$$\underset{\text{2-Butanol}}{CH_3CHOHCH_2CH_3} \xrightarrow[\text{Acid}]{-H_2O} \underset{\text{1-Butene}}{CH_3CH_2CH{=}CH_2} + \underset{\text{2-Butene}}{CH_3CH{=}CHCH_3}$$

3. By the dehalogenation of vicinal dihalogenoalkanes

Alkenes can be prepared by the dehalogenation of vicinal dihalogenoalkanes (compounds in which the halogen atoms are bonded to adjacent carbon atoms). Propene is thus obtained by heating 1,2-dichloropropane with zinc.

$$\underset{\text{1,2-Dichloropropane}}{CH_3CHClCH_2Cl} \xrightarrow{Zn} \underset{\text{Propene}}{CH_3CH{=}CH_2}$$

4. By the partial hydrogenation of alkynes

Alkenes can also be prepared from alkynes. By the partial hydrogenation of alkynes using appropriate reagents and conditions, the *cis*- or *trans*- isomer can be obtained as illustrated in the following examples:

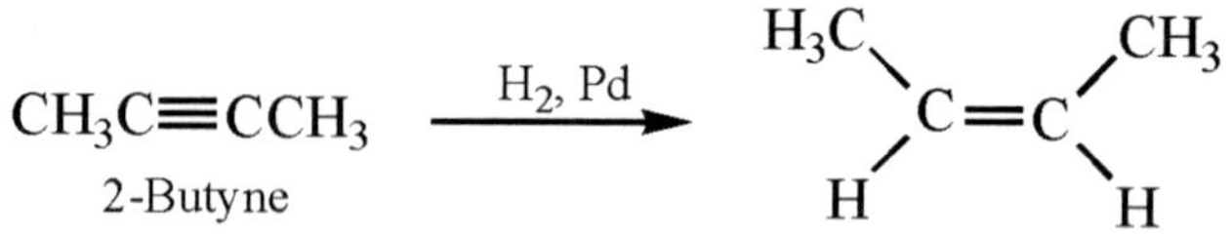

cis-2-butene

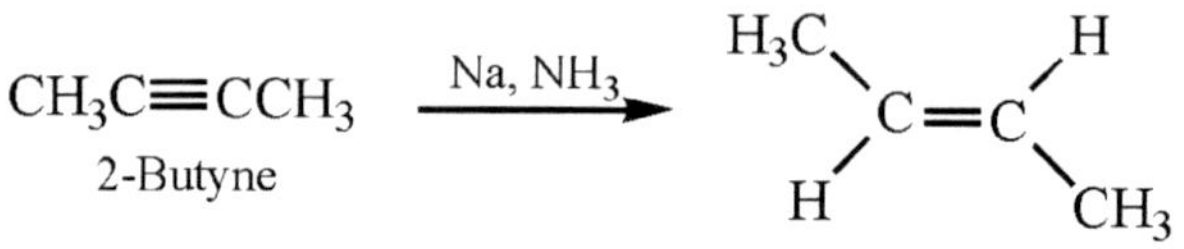

trans-2-butene

► *Key point* – In the partial hydrogenation of alkynes, one molar equivalent of hydrogen is added to the triple bond of an alkyne to give an alkene. When an alkyne is hydrogenated using platinum catalyst, two molar equivalents of hydrogen are generally added to the alkyne to give the corresponding alkane.

6.3 Reactions of alkenes

Alkenes are much more reactive than alkanes. The presence of the π electrons at the double bond makes alkenes susceptible to the attack of electrophiles. Many reagents react with alkenes by the electrophilic addition mechanism. This mechanism is explained in Section 4.5. Some examples of addition reactions are given below.

1. Addition of hydrogen halides

Hydrogen halides add onto alkenes to form halogenoalkanes. When ethene and hydrogen chloride are treated together either as pure reagents or in a polar solvent like ethanoic acid, chloroethane is formed. Propene reacts with hydrogen chloride to give 2-chloropropane. Markovnikov's rule and the reason for the preferential formation of 2-chloropropane are discussed in Section 4.5.

$$\underset{\text{Ethene}}{CH_2{=}CH_2} + HCl \longrightarrow \underset{\text{Chloroethane}}{CH_3CH_2Cl}$$

$$\underset{\text{Propene}}{CH_3CH{=}CH_2} + HCl \longrightarrow \underset{\text{2-Chloropropane}}{CH_3CHClCH_3}$$

If hydrogen chloride is allowed to react with propene in the presence of an alkyl peroxide (R-O-O-R), 1-chloropropane is formed. This is contrary to the Markovnikov's rule of addition of hydrogen halides and is said to be *anti-Markovnikov's addition*.

$$\underset{\text{Propene}}{CH_3CH{=}CH_2} + HCl \xrightarrow{\text{R-O-O-R}} \underset{\text{1-Chloropropane}}{CH_3CH_2CH_2Cl}$$

In this reaction, the mechanism of the addition is different. First, free radicals are formed by the homolysis of the peroxide bond (Step 1). These free radicals abstract hydrogen atoms from H-Cl, producing chlorine atoms (Step 2). In the next step, a chlorine atom adds onto C-1 of propene using an electron from the double bond. A free radical with an electron on C-2 is formed (Step 3), which abstracts a hydrogen from hydrogen chloride to give the product, 1-chloropropane (Step 4).

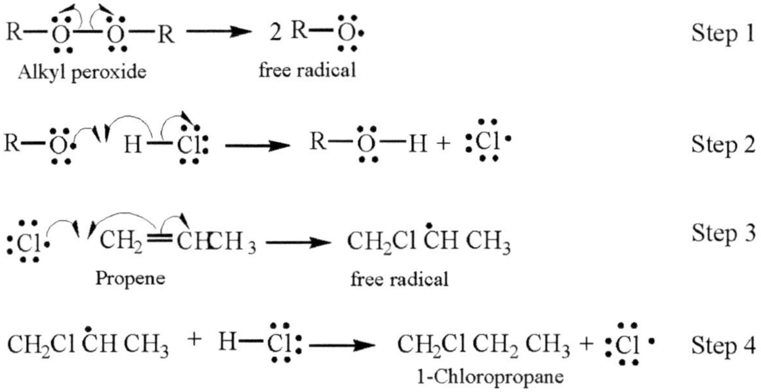

2. Addition of water

Alkenes can be hydrated in the presence of an acid catalyst to form alcohols. When propene adds a molecule of water, 2-propanol is formed.

$$\underset{\text{Propene}}{CH_3\,CH{=}CH_2} + H_2O \xrightarrow{H^+} \underset{\text{2-Propanol}}{CH_3CHOH\,CH_3}$$

The reaction mechanism is similar to the electrophilic addition of hydrogen halides to alkenes. In the first step, propene adds an H^+ ion using a pair of electrons at the double bond to form the intermediate carbocation. This ion then bonds with a water molecule using a lone pair of electrons on the oxygen to form an oxonium ion (an ion with a positive charge on the oxygen), which then eliminates a proton to give the alcohol.

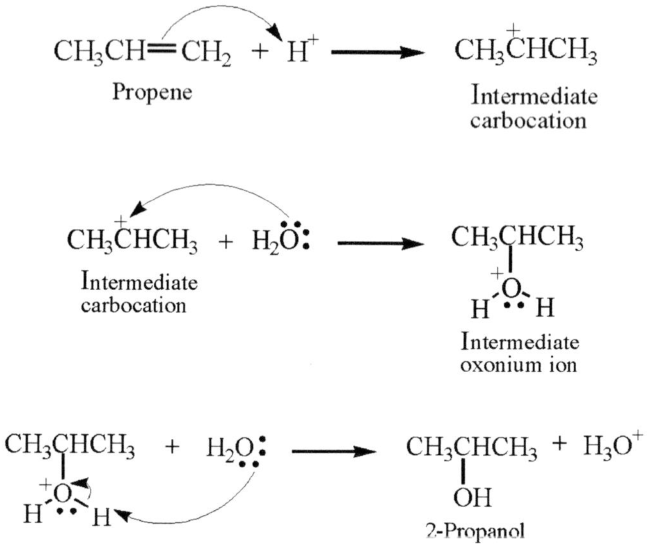

3. Addition of halogens

Chlorine and bromine react with alkenes either as pure reagents or in a solvent like carbon tetrachloride to form dihalides.

$$\underset{\text{2-Butene}}{CH_3\,CH{=}CH\,CH_3} + Br_2 \xrightarrow[CCl_4]{} \underset{\text{2,3-Dibromobutane}}{CH_3\,CHBr\,CHBr\,CH_3}$$

The π electrons at the double bond polarises the bromine molecule (due to the repulsion between the π electrons and the electrons of the Br-Br bond) so that one bromine atom in the molecule develops a positive charge, and the other a negative charge. The alkene bonds with a Br^+ ion to form an intermediate bromonium ion; at the same time a bromide Br^- ion is formed. This can be shown by the following scheme:

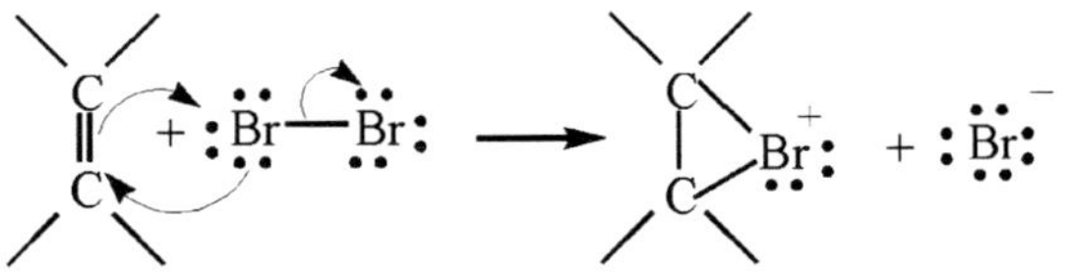

Bromonium ion

The Br^- ion reacts with the intermediate to give the product.

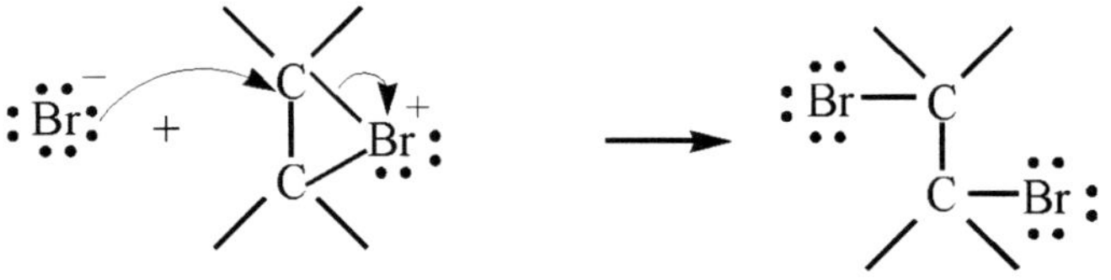

Bromonium ion

4. Addition of hydrogen (a reduction process)

Alkenes add hydrogen in the presence of nickel or palladium catalysts to form alkanes.

$$\underset{\text{Ethene}}{CH_2{=}CH_2} + H_2 \xrightarrow{\text{Ni or Pd}} \underset{\text{Ethane}}{CH_3CH_3}$$

5. Oxidation or hydroxylation

When alkenes are treated with alkaline potassium permanganate, **1,2-diols**, called **glycols** are formed. Here, two hydroxyl groups are added onto the carbon atoms at the double bond. Permanganate is reduced during the reaction to form brown manganese dioxide.

$$\underset{\text{Ethene}}{CH_2{=}CH_2} + 2MnO_4^- + 4\,H_2O \xrightarrow{\text{Base}} \underset{\substack{\text{Ethane-1,2-diol}\\\text{(Ethene glycol)}}}{\begin{array}{cc} CH_2 - & CH_2 \\ | & | \\ OH & OH \end{array}} + 2MnO_2 + 2OH^-$$

A similar reaction takes place when an alkene is treated with osmium tetroxide (OsO_4), followed by sodium sulphite (Na_2SO_3) solution. Both of these reactions take place with the formation of an intermediate cyclic addition product, which hydrolyses to give the diol. Cyclohexene on oxidation gives cis-1,2-cyclohexanediol.

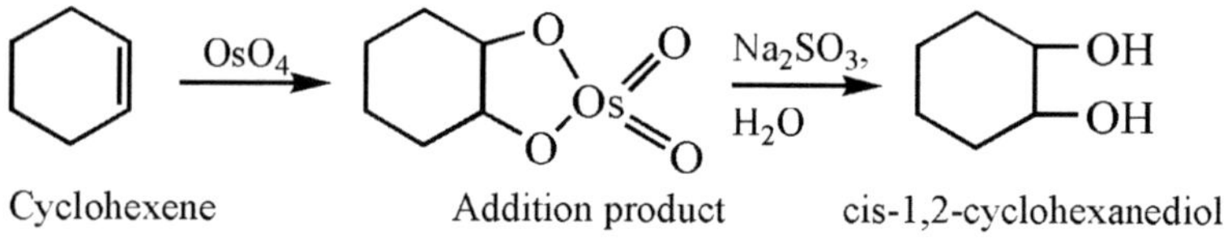

6. Polymerisation

Alkenes undergo polymerisation reactions, in the presence of catalysts like peroxides to produce polymers. Ethene polymerises to give polyethene or polythene.

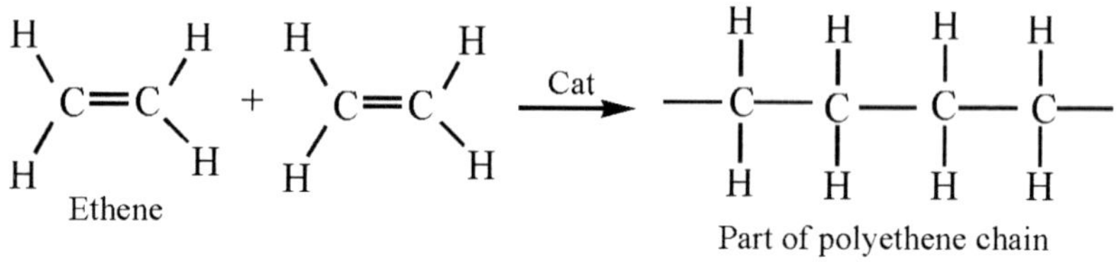

Benzoyl peroxide acts as a catalyst in this reaction. It breaks into two free radicals (1). These are very reactive and initiate the reaction by taking an electron from the π electrons of ethene and bonding to it, thus forming a free radical (2). This reacts with another molecule of ethene forming a bond between the two molecules (3) and setting up a chain reaction. Thousands of ethene molecules can thus link together to form large molecules of very high molecular weight. The reaction terminates when free radicals unite to form bonds (5).

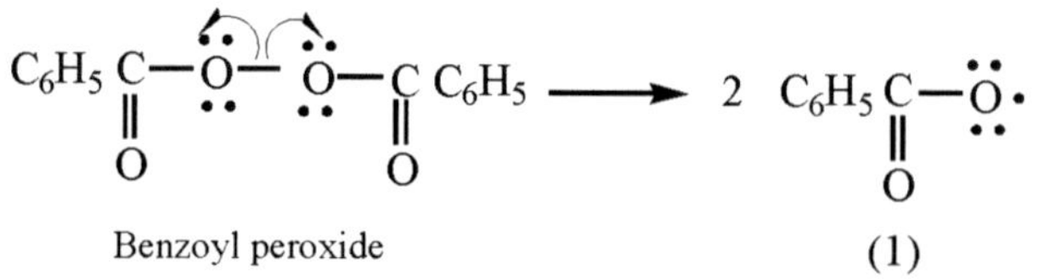

(1)

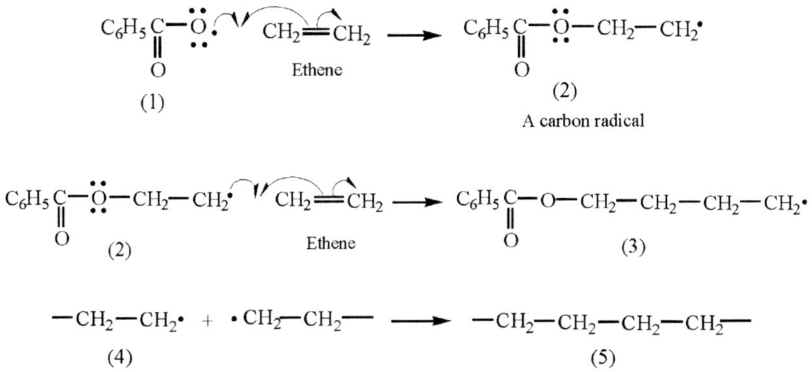

6.4 Alkynes

The general formula of alkynes is C_nH_{2n-2}. The simplest alkyne is ethyne (acetylene). It is a linear molecule with three covalent bonds (triple bond) between the carbon atoms, one of which is a σ bond and the other two π bonds. The four π electrons delocalize to form a cylindrically shaped electron cloud around the σ bond (see Section 1.4 for the *sp* hybridisation of carbon and the formation of an ethyne molecule).

The three carbon alkyne is propyne (methylacetylene). Ethyne and propyne have only one isomer each, of structural formulae:

$$H-C\equiv C-H \qquad CH_3C\equiv C-H$$

Ethyne Propyne

Butyne has two constitutional isomers; 1-butyne and 2-butyne.

$$CH_3CH_2C\equiv C-H \qquad CH_3C\equiv C\,CH_3$$

1-Butyne 2-Butyne

An alkyne which has the triple bond between C-1 and C-2 is called a **terminal alkyne**. Propyne, 1-butyne and 1-pentyne are examples of terminal alkynes.

$$CH_3C\equiv C-H \qquad CH_3CH_2C\equiv C-H \qquad CH_3CH_2CH_2C\equiv C-H$$

Propyne 1-Butyne 1-Pentyne

6.5 Preparing alkynes

1. By the dehydrohalogenation of vicinal dihalides

A convenient method to prepare an alkyne is by the dehydrohalogenation of a vicinal dihalide with a strong base like sodamide ($NaNH_2$) in liquid ammonia. 2-Butyne can be prepared from 2,3-dibromobutane as illustrated by the following scheme:

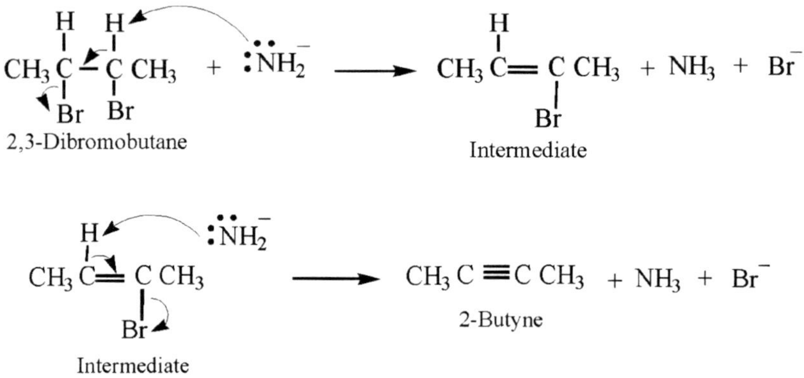

2. A higher alkyne from a lower terminal alkyne

A lower terminal alkyne can be converted to an alkynide ion which reacts with an alkyl halide to give a higher alkyne. For example, 1-pentyne can be prepared from ethyne. When ethyne is treated with sodamide in liquid ammonia, sodium ethynide is formed, which reacts with 1-bromopropane to form 1-pentyne.

$$\underset{\text{Ethyne}}{H-C\equiv C-H} + \underset{\text{Sodamide}}{NaNH_2} \xrightarrow{NH_3} \underset{\text{Sodium ethynide}}{H-C\equiv C^-Na^+} + NH_3$$

$$H-C\equiv C^-Na^+ + \underset{\text{1-Bromopropane}}{CH_3CH_2CH_2Br} \xrightarrow{NH_3} \underset{\text{1-Pentyne}}{H-C\equiv C\,CH_2CH_2CH_3} + NaBr$$

6.6 Reactions of alkynes

1. Addition reactions

Alkynes undergo addition reactions with hydrogen, hydrogen halides and halogens, the mechanism of the addition reactions being similar to that of alkenes. A molecule of alkyne adds one diatomic molecule to give an alkene which adds a second molecule to give a saturated compound. For example, ethyne reacts with a molecule of HCl to give chloroethene which in turn adds a second molecule of HCl to give 1,1-dichloroethane.

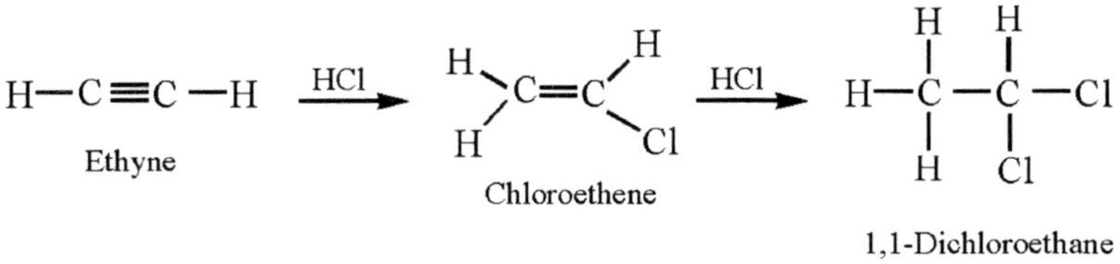

2. Substitution reactions of terminal alkynes

Acetylenic hydrogen atoms are acidic. These can be substituted by sodium, silver and copper (1).

When propyne is treated with an ammoniacal solution containing silver ion or copper (1) ion, precipitates of the metal substituted products are formed.

$$\underset{\text{Propyne}}{CH_3C \equiv C-H} + Ag(NH_3)_2^+ + OH^- \longrightarrow \underset{\text{Precipitate}}{CH_3C \equiv CAg} + 2NH_3 + H_2O$$

Alkynes with sodamide in liquid ammonia give sodium derivative of the alkyne.

$$\underset{\text{Propyne}}{CH_3C \equiv C-H} + NaNH_2 \xrightarrow{NH_3} CH_3C \equiv C^- Na^+ + NH_3$$

6.7 Arenes

Benzene is the parent hydrocarbon of **arenes** or **aromatic compounds**. Benzene, C_6H_6, has a ring of six carbon atoms and a hydrogen atom attached to each carbon. The carbon atoms are often represented as being bonded by alternating single and double bonds. However, this is an over-simplification of the structure of benzene. The π electrons in the double bonds are not localised. They are delocalized, forming an electron cloud residing above and below the plane of the carbon ring (Section 2.11).

Substituted benzenes are named as derivatives of benzene. For example, the compound in which a hydrogen atom of benzene is substituted by a methyl group, is called methylbenzene. It is often called by its trivial name, toluene. The IUPAC nomenclature retains many trivial names. In the following examples of monosubstituted benzenes, trivial names are given in brackets.

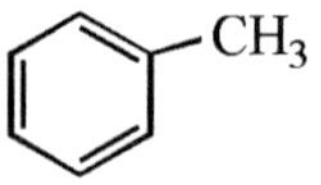

Methylbenzene (Toluene)

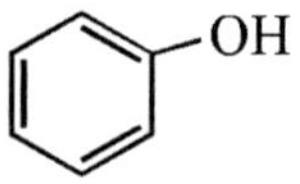

Hydroxybenzene (Phenol)

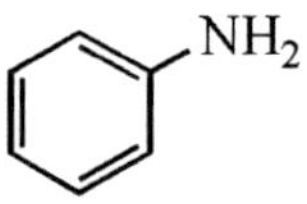

Aminobenzene (Aniline)

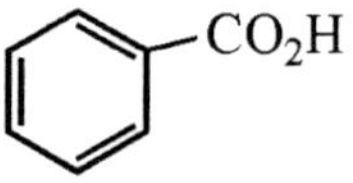

Benzoic acid

For disubstituted aromatic compounds a numbering system or *ortho*- (1,2- position), *meta*- (1,3- position) and *para*- (1,4- position) prefixes are used. If there are more than two substituents, numbers are used to locate their positions.

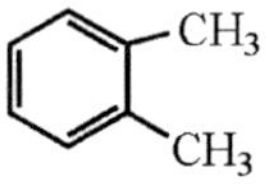

1,2-Dimethylbenzene (o-Xylene)

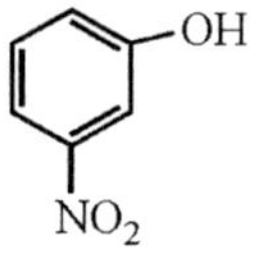

3-Nitrophenol or m-Nitrophenol

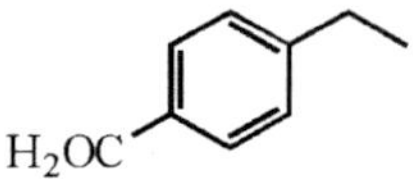

4-Ethylbenzoic acid or p-Ethylbenzoic acid

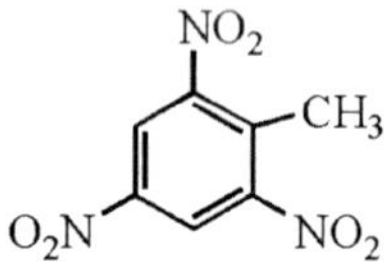

2,4,6-Trinitrotoluene

The C_6H_5- group is called a phenyl group (Ph-) and $C_6H_5CH_2$- a benzyl group.

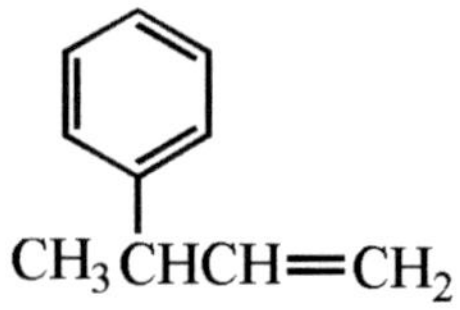

3-Phenyl-1-butene

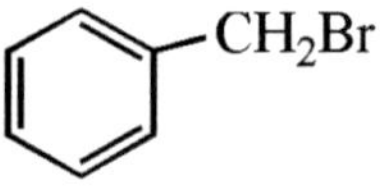

Benzyl bromide

6.8 Reactions of benzene

1. Hydrogenation of benzene and alkyl benzenes

Benzene and alkylbenzenes are hydrogenated with hydrogen in the presence a nickel or platinum catalyst under high pressure to form cycloalkanes.

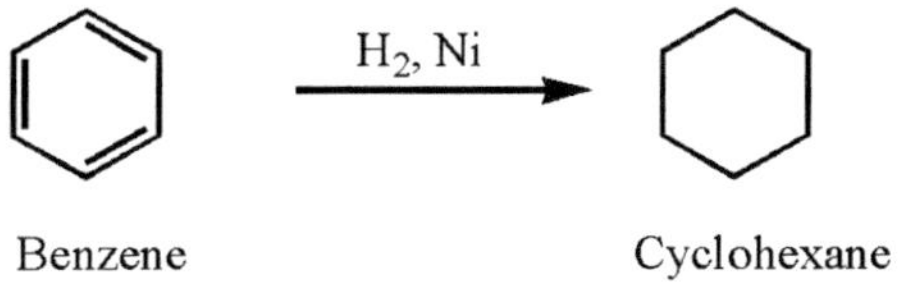

2. Electrophilic substitution reaction

Benzene readily undergoes substitution reaction, in spite of its high unsaturation. The fact that the benzene ring tends to retain its stable π electron cloud system during reactions, is due to a property called **aromatic stabilisation**.

Electrophiles are attracted by the negative electron cloud provided by the π electrons. An electrophile forms a covalent bond with one of the carbon atoms of benzene using a pair of π electrons. This breaks up the stable electron cloud system, but the intermediate positive ion formed is stabilised to some extent due to the formation of a resonance hybrid consisting of the three structures (I-III) as shown below. In the next step of the reaction, the intermediate eliminates a proton from the carbon atom attached to the electrophile. The electron pair shared between that carbon and the proton joins the delocalized electron system to form a product which retains the stable structure of benzene. This mechanism is explained below using the symbol E^+ for the electrophile.

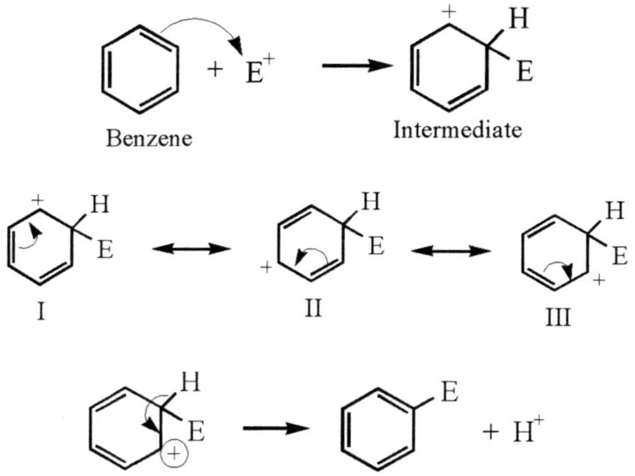

The following are examples of electrophilic substitution.

1. Halogenation

Benzene reacts with chlorine in the presence of iron, iron (III) chloride or aluminium chloride as catalyst, to give chlorobenzene and hydrogen chloride. In this reaction, the electrophile is Cl^+. The reaction can be shown as:

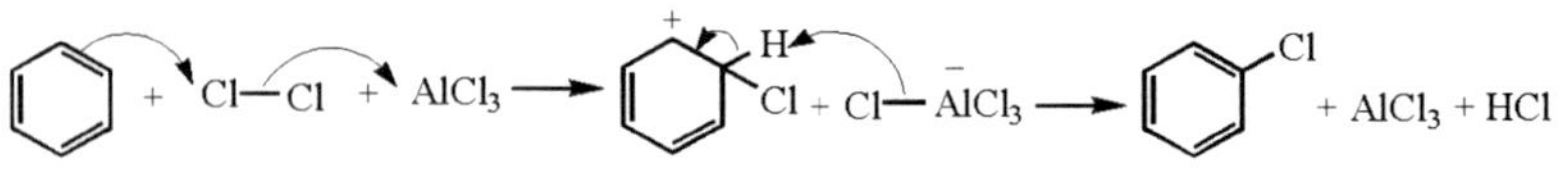

The overall reaction is:

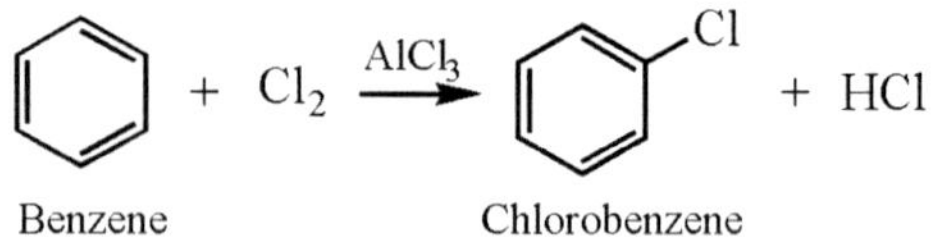

Bromine reacts with benzene under similar conditions, but at higher temperatures to form bromobenzene.

2. Nitration

Benzene reacts with concentrated nitric acid in the presence of concentrated sulphuric acid catalyst to form nitrobenzene. This reaction takes place by a mechanism similar to halogenation. The electrophile nitronium ion, NO_2^+, is formed by the reaction between nitric acid and sulphuric acid.

$$HNO_3 + 2H_2SO_4 \longrightarrow NO_2^+ + 2HSO_4^- + H_3O^+$$

The overall reaction is:

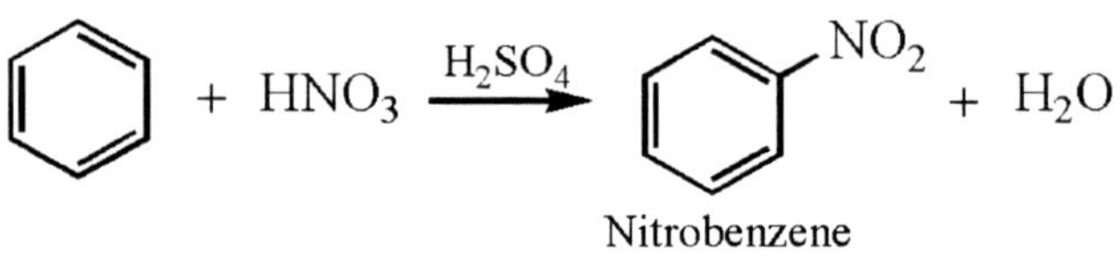

3. Sulphonation

Benzene reacts with fuming sulphuric acid (sulphuric acid which contains dissolved sulphur trioxide) to produce benzenesulphonic acid.

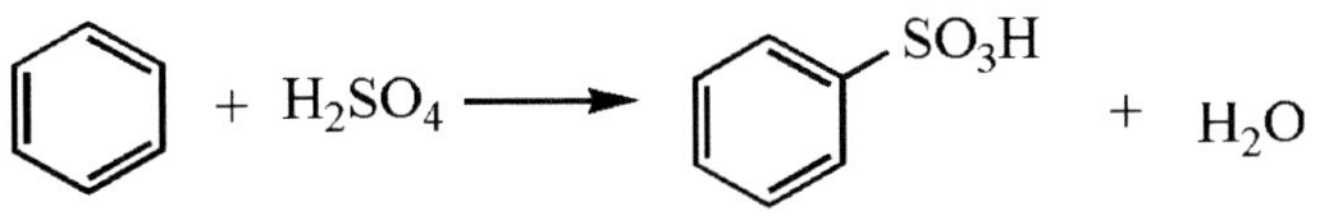

Benzenesulphonic acid

4. Friedel-Crafts alkylation

When benzene is treated with an alkyl chloride in the presence of a catalyst like aluminium chloride, alkyl benzene is formed. Thus, benzene reacts with chloroethane to give rise to ethylbenzene.

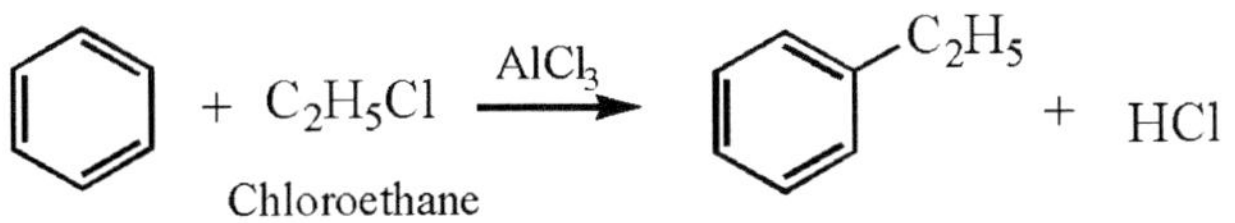

Chloroethane

Ethylbenzene

6.9 Effect of a substituent on further substitution

Toluene, $C_6H_5CH_3$, is more reactive than benzene to electrophilic reagents. A group like methyl group, attached to the benzene ring, which increases the reactivity of the ring, is called an **activating group**. Some other activating groups are -OH, $-OCH_3$, $-NH_2$ and C_6H_5-. Groups which lower the reactivity of the ring are called **deactivating groups**. $-NO_2$, $-CO_2H$, $-CO_2R$, $-CF_3$, $-CCl_3$, $-SO_3H$ and -Cl are some examples.

Substituents which increase the reactivity of the benzene ring are **electron releasing groups**. They stabilize the intermediate carbocation thus increasing its reactivity. On the other hand, **electron withdrawing groups**, such as $-NO_2$, $-CO_2H$ and -F, lower the electron density of the ring thus lowering the reactivity with electrophiles.

When toluene is nitrated with concentrated nitric acid and sulphuric acid, a mixture of o- and p-nitrotoluenes is formed. But when nitrobenzene is nitrated under similar conditions, m-dinitrobenzene is the major product.

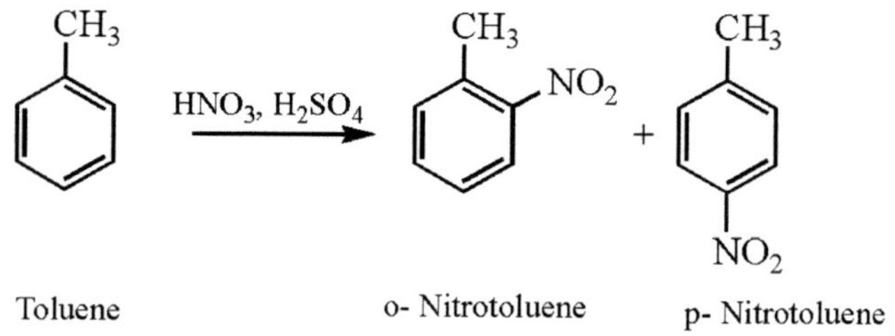

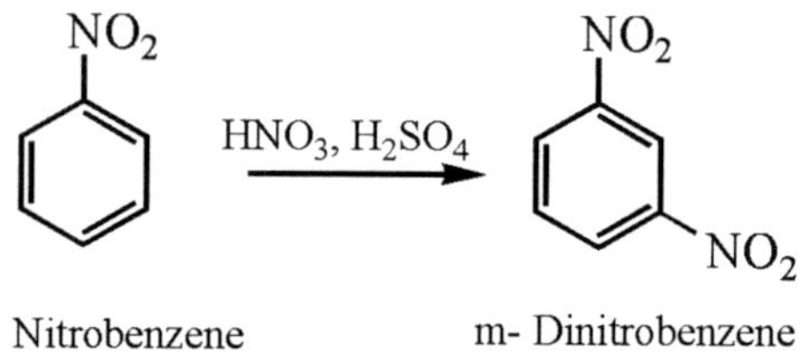

The influence of the methyl group substituent is that it is **ortho- para-directing**. In the first step of the reaction, the nitronium ion (NO_2^+) can attack the ortho-, meta- or para- position to form the intermediate carbocation. The ortho- and para- substituted intermediate carbocations are more stabilized (than the meta) by the electron-donating inductive effect of the methyl group, since they have a structure containing the positive charge on the carbon atom attached to the methyl group (highlighted in the scheme).

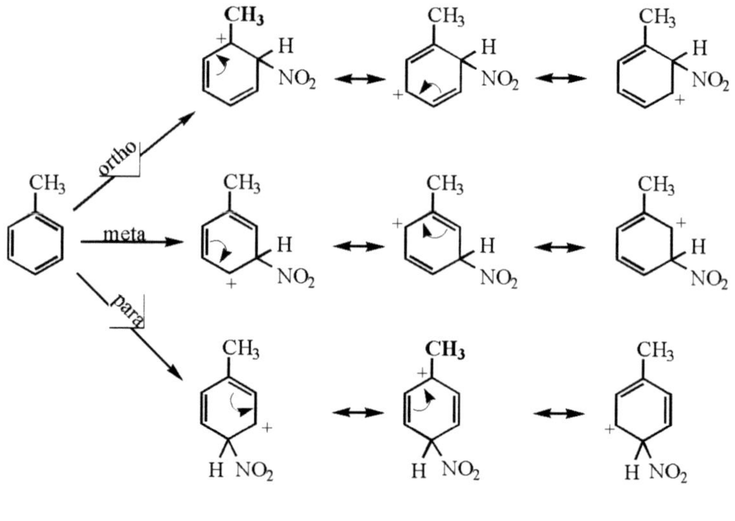

The nitro group is **meta-directing**. The electron withdrawing inductive effect of the nitro group makes the ortho- and para-substituted intermediate carbocations less stable than the meta- one, and so the electrophilic substitution takes place at the least deactivated meta-position. This is summarised in Table 6.1.

ortho-para directing	
Strongly activating	$-NH_2$ -NHR $-NR_2$ -OH
Moderately activating	-NHCOR $-OCH_3$ -OR -OCOR
Weakly activating	$-CH_3$ -R $-C_6H_5$
Weakly deactivating	-F -Cl -Br -I
meta directing	
Moderately deactivating	$-CO_2H$ $-CO_2R$ -CHO -COR $-CONH_2$
Strongly deactivating	$-NO_2$ $-CF_3$ $-CCl_3$ $-SO_3H$

Table 6.1. The effect of a substituent on further substitution.

Tutorial: helping you learn

Progress questions

1. 1-Hexene is treated with bromine to give a colourless product.
   (a) Write an equation for the reaction.
   (b) Give the IUPAC name of the product.
   (c) Give the mechanism of the reaction.
   (d) What product/s are formed if bromine is added to a suspension of 1-hexene in aqueous sodium chloride? Give your reason(s).

2. Give the formula of the carbocations that can be formed by the addition of H^+ to each of the following alkenes. Indicate which is more stable in each case:

(a) $CH_3\,CH{=}CH_2$

(b) $CH_3\underset{\displaystyle CH_3}{\underset{|}{C}}{=}CHCH_2CH_3$

(c) $CH_3CH{=}\underset{\displaystyle CH_3}{\underset{|}{C}}\,CH_2CH_3$

(d) 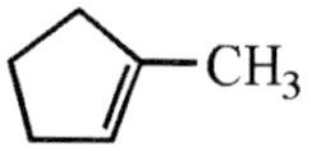

3. Give the reagents and reaction conditions for the following conversions.

(a) 

2-Butyne → *cis*-2-Butene

(b) 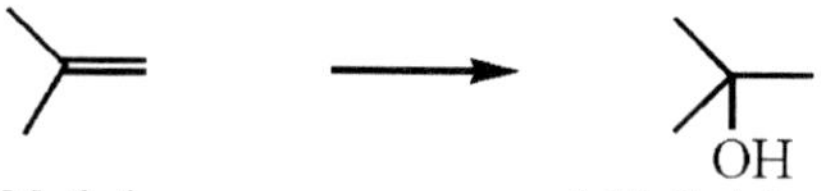

2-Methylpropene → 2-Methyl-2-propanol

(c) 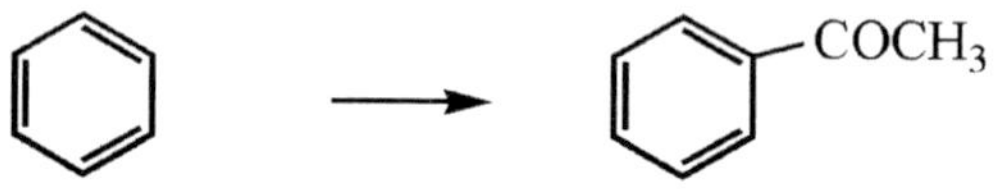

Benzene → Acetophenone

(d) $CH_3C{\equiv}CH \longrightarrow CH_3C{\equiv}CCH_2CH_3$

Propyne → 2-Pentyne

4. Name and discuss the mechanism of the reactions in the following cases.

(a) 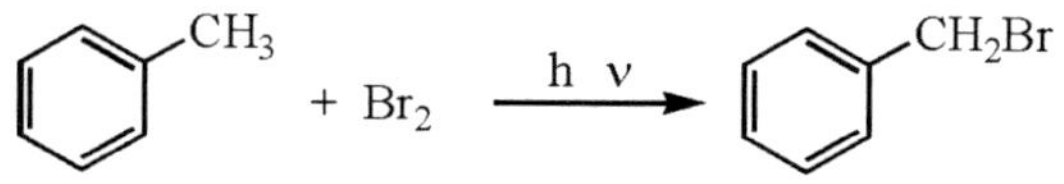

(b) 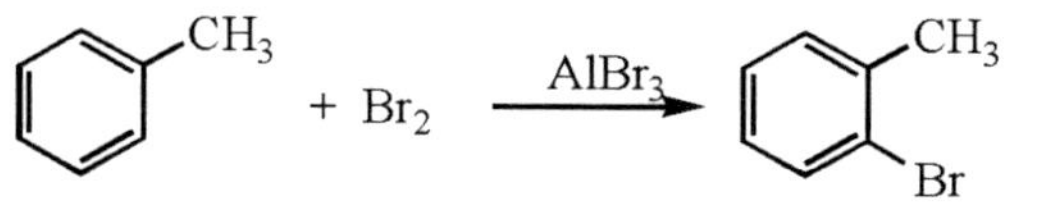

5. Show the steps involved in the conversion of ethylbenzene to phenylacetylene.

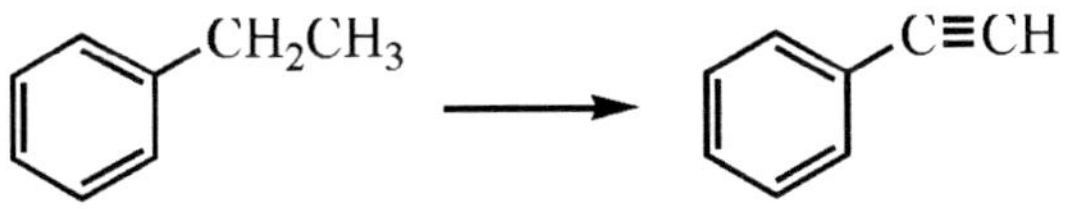

Ethylbenzene Phenylacetylene

Discussion point

We have seen in Section 6.9 that an electron-releasing substituent like methyl group on the benzene ring directs further electrophilic substitution to ortho-para positions and an electron-withdrawing substituent like nitro group, to the meta position. Account for this pattern of substitution.

Practical assignment

Investigate the chemistry of benzene side-chain oxidation and halogenation reactions.

Study tips

1. Compare the molecular formulae of an alkene and a cycloalkane of same number of carbon atoms. (Example: hexene, C_6H_{12} and cyclohexane, C_6H_{12}) They have the same molecular formula and have two hydrogen atoms less than the corresponding alkane, hexane, C_6H_{14}. A double bond in hexene or a ring in cyclohexane is equivalent to two hydrogen atoms.

2. A triple bond is equivalent to four hydrogen atoms or two double bonds or a ring and a double bond.

3. Organic chemists can get some information about the structure of a compound from its molecular formula by finding the difference between the number of hydrogen atoms in the compound under consideration and the number of hydrogen atoms in the corresponding alkane.

7

Alcohols and Phenols

One-minute summary – The functional group in alcohols and phenols is the hydroxyl (-OH) group. This is bonded to an alkyl group in alcohols and to an aryl group in phenols. The presence of the polar hydroxyl group accounts for their solubility in water, and their high melting and boiling points. Though alcohols can act as weak acids and weak bases, they are generally considered as neutral. Phenols are weak acids. In this chapter we will discuss:

- hydrogen bonding in alcohols and phenols
- acidity and acid constant
- comparison of the acidities of alcohols and phenols
- nucleophilic substitution and elimination reactions of alcohols
- oxidation reactions of alcohols
- electrophilic substitution reactions of phenols.

7.1 Alcohols

Alcohols are compounds containing a hydroxyl (-OH) group attached to an sp^3 hybridised carbon atom. They can be considered as compounds formed by substituting a hydrogen atom of water by an alkyl group.

Alcohols which contain one hydroxyl group are called **monohydric alcohols**. Monohydric alcohols form a homologous series with the general formula $C_nH_{2n+2}O$ or $C_nH_{2n+1}OH$. Alcohols are classified as primary (1°), secondary (2°) and tertiary (3°) depending on the position of the -OH group in the molecule. A **primary alcohol** has a -CH_2OH group, i.e. the alcoholic carbon is bonded to at least two hydrogen atoms. A **secondary alcohol** has a -CHOH group, i.e. carbon bonded to two alkyl groups and a **tertiary alcohol** has a -COH group, bonded to three alkyl groups. Examples of the three kinds of alcohols are given below.

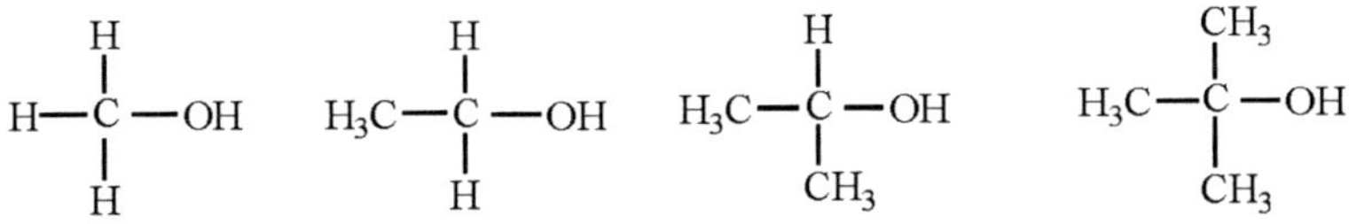

HCH_2OH	CH_3CH_2OH	CH_3-CHOH (with CH_3 on C)	2-Methyl-2-propanol
Methanol	Ethanol	2-Propanol	(tertiary)
(primary)	(primary)	(secondary)	

Polyhydric alcohols are those which contain more than one hydroxyl group. Alcohols which contain two hydroxyl groups are called **dihydric alcohols** or **diols,** and those which contain three are called **trihydric alcohols** or **triols**. Ethane-1,2-diol ($HOCH_2CH_2OH$) and propane-1,2,3-triol ($HOCH_2CHOHCH_2OH$) are examples of polyhydric alcohols.

Like water, alcohols are polar covalent compounds. The hydroxyl group is bonded to an sp^3 hybridised carbon atom and the more electronegative oxygen attracts shared electron pairs towards it. Thus, the oxygen atom of the alcoholic group carries a δ- charge, and the carbon and hydrogen atoms a δ+ charge; both the C-O and O-H bonds are polar.

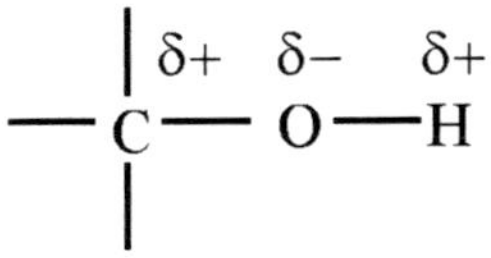

Molecules of alcohols attract one another through the negative oxygen of one molecule and the positive hydrogen of another molecule. This attractive force is called **hydrogen bonding**. In an alcohol (like in water), molecules are hydrogen bonded. Hydrogen bonds are stronger than other intermolecular bonds such as van der Waals forces which are normally present in organic compounds. This accounts for the high boiling points of alcohols compared to those of hydrocarbons, halogenoalkanes or ethers of similar molecular weight.

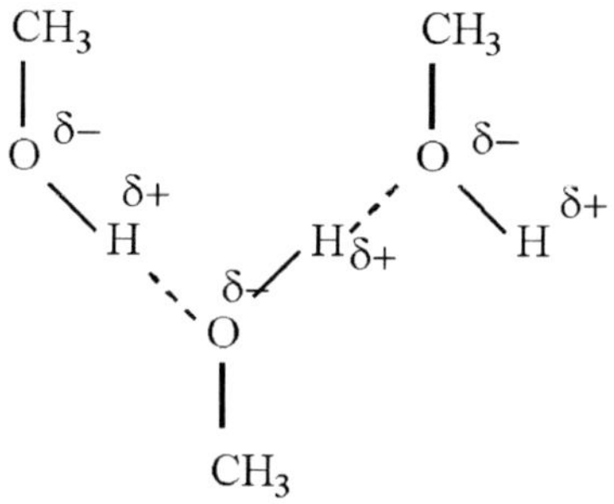

Hydrogen bonding between methanol molecules, shown by dotted lines

If there is more than one hydroxyl group in a molecule of an alcohol, its boiling point is higher than that of a monohydric alcohol of similar molecular weight, as expected.

Compare the boiling points (b.pt) of the following compounds:

Compound	*Structural formula*	*Molecular wt*	*b.pt (°C)*
1-Butanol	$CH_3CH_2CH_2CH_2OH$	74	117
Pentane	$CH_3CH_2CH_2CH_2CH_3$	72	36
Diethylether	$CH_3CH_2OCH_2CH_3$	74	35
Chloroethane	CH_3CH_2Cl	64.5	12.5
1-Pentanol	$CH_3CH_2CH_2CH_2CH_2OH$	88	138
1,4-Butanediol	$HOCH_2CH_2CH_2CH_2OH$	90	230

Table 7.1. The properties of alcohols.

The presence of the polar hydroxyl group accounts for the solubility of the lower alcohols in water, since molecules of water and alcohol form intermolecular hydrogen bonds. The solubility in water decreases as the size of the carbon chain increases.

7.2 Acidity of alcohols

Acidity and acid constant

Alcohols behave as weak acids. A study of the Bronsted-Lowry theory of acids and bases and the acid constant will help us to understand and compare the acid nature of alcohols and other compounds. A **Bronsted-Lowry acid** is a proton donor and a **Bronsted-Lowry base** is a proton acceptor. An acid dissociates in water, giving a proton to water. For the dissociation of an acid HA in water, the following equilibrium equation can be written:

$$\underset{\text{Acid}}{HA} + \underset{\text{Base}}{H_2O} \rightleftharpoons \underset{\text{Base}}{A^-} + \underset{\text{Acid}}{H_3O^+}$$

- *Key point* – When HA loses a proton, it becomes A^-, a base. In the reverse reaction, A^- accepts a proton. Such an acid-base pair is called a **conjugate acid-base pair**. A^- is the conjugate base of acid HA. Another conjugate acid-base pair is H_3O^+-H_2O.

The equilibrium constant for the dissociation of an acid is called the **acid constant** and is given by the expression,

$$K_a = \frac{[A^-][H_3O^+]}{[HA]}$$

- *Key point* – H_2O is in large excess in a reaction mixture, so $[H_2O]$ is regarded as constant and does not appear in the equation.

The value of K_a is constant for an acid at a constant temperature. The higher the value of K_a for an acid, the greater its dissociation, and the stronger the acid. HCl is a strong acid, its K_a value being 10^7. Ethanoic acid is a weak acid, its K_a being 1.8×10^{-5}.

It is convenient to express acid strengths as pK_a values which are negative logarithms of K_a values:

$$pK_a = -\log K_a$$

The higher the pK_a value, the lower the acid strength, or the weaker the acid. The pK_a value of HCl is -7 and that of ethanoic acid is 4.75. Note that the conjugate base of a strong acid is weak and the conjugate base of a weak acid is strong. By the same token, ethanoate is a stronger base than chloride. The pK_a values of some acids are given below for comparison.

Compound	*Formula*	pK_a
Ammonia	NH_3	33
Ethanol	C_2H_5OH	16
Water	H_2O	15.7
Phenol	C_6H_5OH	9.95
Ethanoic acid	CH_3CO_2H	4.75
Benzoic acid	$C_6H_5CO_2H$	4.2
Hydrochloric acid	HCl	-7

Table 7.2. pK_a values.

Alcohols as acids

Alcohols are weak acids. Alcohols undergo a small degree of dissociation in water as shown by the equation,

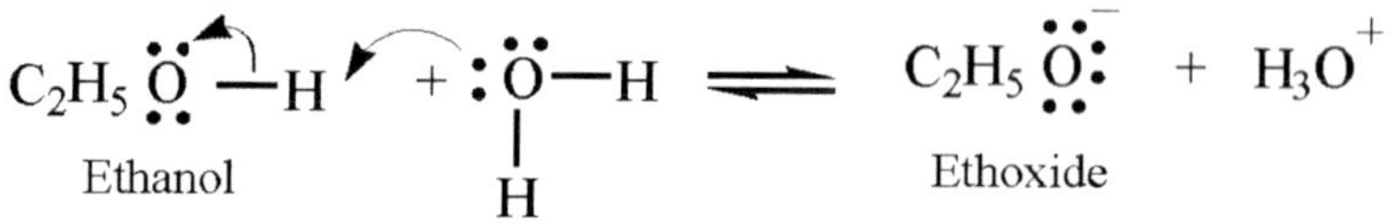

The acid constant, K_a, of ethanol is 10^{-16}. The small K_a value shows that ethanol is a very weak acid. The pK_a of ethanol is 16. It is a stronger acid than ammonia, but is weaker than water and ethanoic acid. (Note the pK_a values of ammonia, ethanol, water and ethanoic acid in Table 7.2.)

The acid nature of alcohols is shown by their reaction with strong bases like hydrides. For example, sodium hydride reacts with ethanol to give sodium ethoxide. Alcohols also react with strong electropositive metals like sodium, potassium and magnesium to give alkoxides and hydrogen gas.

$$C_2H_5OH + NaH \longrightarrow C_2H_5O^- Na^+ + H_2$$

Ethanol — Sodium hydride — Sodium ethoxide

$$2\ C_2H_5OH + 2\ Na \longrightarrow 2\ C_2H_5O^- Na^+ + H_2$$

Ethanol — Sodium ethoxide

7.3 Alcohols as bases

Alcohols can also act as weak bases, and can accept protons from strong acids:

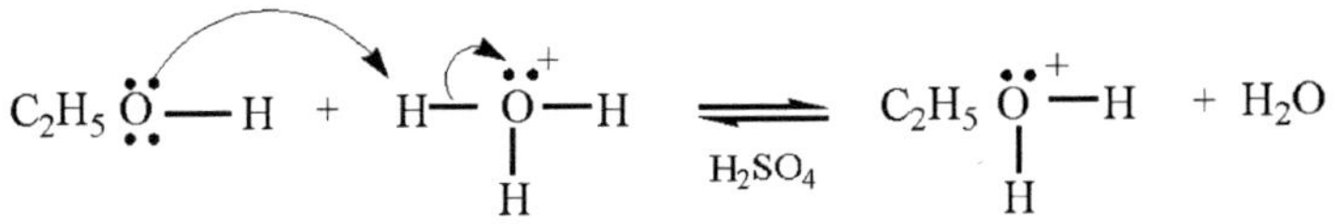

An oxonium ion

7.4 Conversion of alcohols to halogenoalkanes

Alcohols react with hydrogen chloride, hydrogen bromide, hydrogen iodide, phosphorus pentachloride, phosphorus tribromide and thionyl chloride to give halogenoalkanes. Study the following examples:

$$\underset{\text{Ethanol}}{CH_3CH_2OH} + HCl \xrightarrow[\text{ZnCl}_2,\ \text{heat}]{} \underset{\text{Chloroethane}}{CH_3CH_2Cl} + H_2O$$

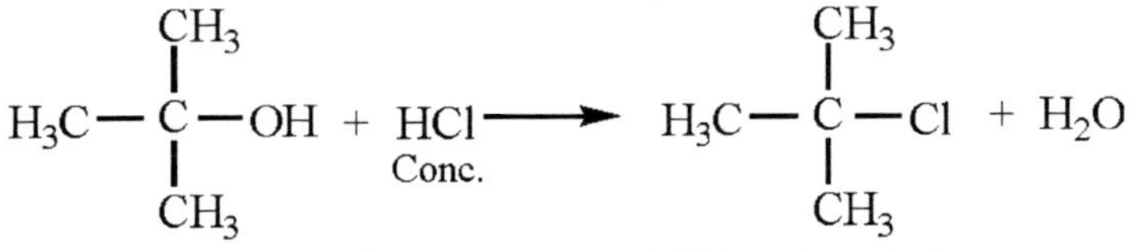

2-Methyl-2-propanol → 2-Chloro-2-methylpropane

$$3\ \underset{\text{Ethanol}}{C_2H_5\text{-}OH} + PBr_3 \xrightarrow[\text{heat}]{} 3\ \underset{\text{Bromoethane}}{C_2H_5\text{-}Br} + H_3PO_3$$

The reactivity of hydrogen halides increases in the following order:

$$HCl < HBr < HI.$$

The reactivity of the alcohols increases in the order

primary < secondary < tertiary.

The formation of halogenoalkanes from alcohols is by nucleophilic substitution reaction. Primary alcohols react with the nucleophiles by an S_N2 mechanism, and tertiary alcohols by an S_N1 mechanism (as shown by the schemes below). Secondary alcohols react by either mechanism depending on the reaction conditions.

The first step in the reaction between ethanol and concentrated hydrochloric acid is the protonation of the alcohol; this step is fast and reversible. In the second step, the nucleophile attacks the positive carbon, forming a bond with it. At the same time the C-O bond breaks and a molecule of water is eliminated. This step is slow and is thus the rate-determining step. Since two reactant species are involved in this step, it is the S_N2 mechanism.

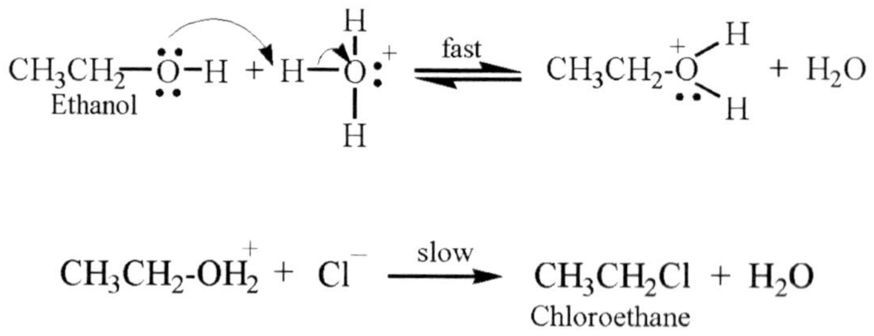

On the other hand, a protonated tertiary alcohol eliminates a molecule of water to form a stable cation. This is a slow reaction, involving only one reactant species; the S_N1 mechanism operates here. The carbocation adds the nucleophile to form the product.

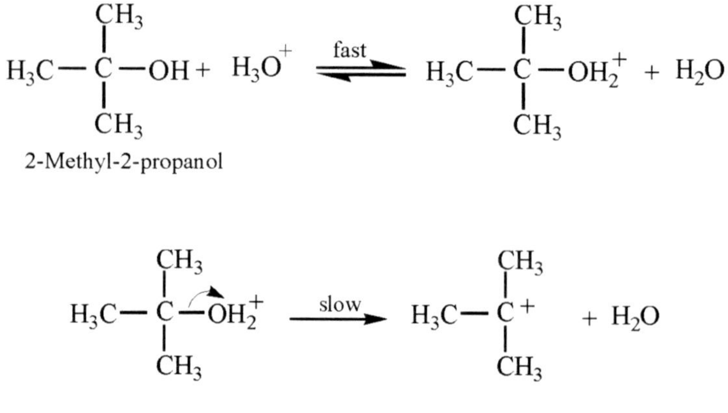

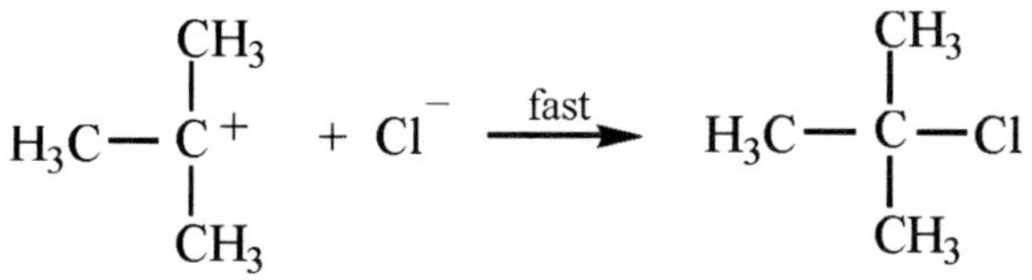

2-Chloro-2-methylpropane

7.5 Dehydration of alcohols

Alcohols can be dehydrated to alkenes by treating them with a mineral acid to act as a dehydrating agent. Common dehydrating agents are 85% phosphoric acid or concentrated sulphuric acid, at temperatures from 100 to 200°C. Tertiary alcohols can be dehydrated in milder conditions. The mechanism of dehydration reactions is explained in Section 4.11.

$$\underset{\text{Ethanol}}{CH_3CH_2OH} \xrightarrow[\text{heat}]{\text{Conc. } H_2SO_4} \underset{\text{Ethene}}{CH_2{=}CH_2} + H_2O$$

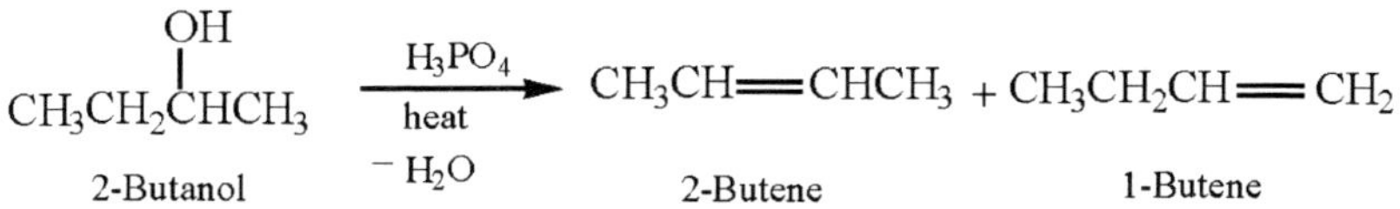

2-Butanol 2-Butene 1-Butene

7.6 Oxidation of alcohols

A primary alcohol can be oxidised to an aldehyde or to a carboxylic acid depending upon the oxidising agent and the reaction conditions. Oxidising agents commonly used include chromium trioxide (CrO_3), potassium dichromate ($K_2Cr_2O_7$) and potassium permanganate ($KMnO_4$) in aqueous sulphuric acid.

$$\underset{\text{1-Butanol}}{CH_3CH_2CH_2CH_2OH} \xrightarrow[\text{aq. } H_2SO_4]{CrO_3} \underset{\text{Butanal}}{CH_3CH_2CH_2CHO} \xrightarrow[\text{aq. } H_2SO_4]{CrO_3} \underset{\text{Butanoic acid}}{CH_3CH_2CH_2CO_2H}$$

There are other oxidising agents which are used for selective oxidation. Pyridinium chlorochromate (PCC), ($C_5H_5NH^+ClCrO_3^-$), pyridinium dichromate (PDC), $[(C_5H_5NH^+)_2Cr_2O_7^{2-}]$, Chromium trioxide-pyridine complex (Collin's reagent), $[(C_5H_5N)_2CrO_3]$ are some of the oxidising agents used in synthesis for carrying out oxidation reactions. A primary alcohol can be selectively oxidised to an aldehyde using one of these oxidising agents.

A secondary alcohol can be oxidised by one of the above oxidising agents to a ketone. A tertiary alcohol can not be oxidised.

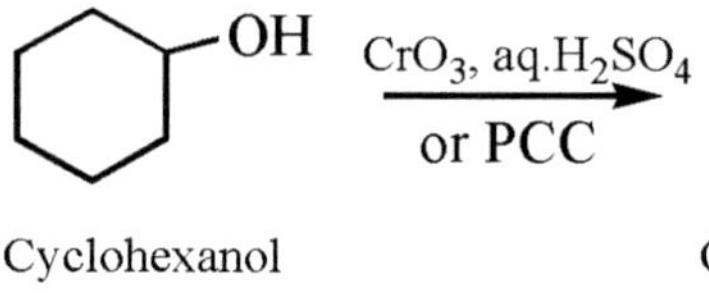

7.7 Phenols

Nomenclature and properties

Phenols are aromatic compounds that have hydroxyl groups attached directly to the benzene ring. We have already introduced phenols in Chapter 2 (Section 2.14). The formulae and the names of some phenols are given below.

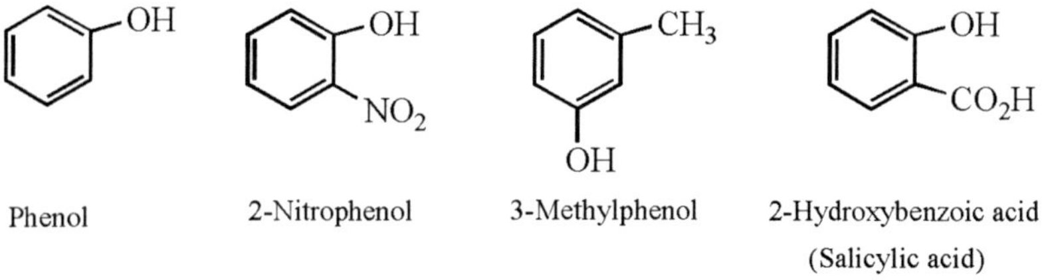

Since phenols contain a hydroxyl group, the molecules form fairly strong intermolecular hydrogen bonds, thus having high boiling points. For example, the boiling point of phenol (mol.weight, 94) is 182°C, whereas the boiling point of toluene, a hydrocarbon of similar molecular weight (mol.weight, 92) is 111°C. Phenols are also partially soluble in water due to hydrogen bonding with water molecules.

Phenols as acids

Phenols are weak acids. They are more acidic than alcohols. They react with active metals and with alkalis to form metal phenoxides. Metal phenoxides are ionic and are soluble in water, hence a phenol dissolves in an aqueous alkali. Alcohols do not react with alkalis.

$$\underset{\text{Phenol}}{2\,C_6H_5OH} + 2\,Na \longrightarrow \underset{\text{Sodium phenoxide}}{2\,C_6H_5O^-Na^+} + H_2$$

$$\underset{\text{Phenol}}{C_6H_5OH\,(l)} + NaOH\,(aq) \longrightarrow \underset{\text{Sodium phenoxide}}{C_6H_5O^-Na^+(aq)} + H_2O$$

The pK_a value of phenol is 9.9 which is less than that of cyclohexanol, an alcohol. Thus, phenol is a stronger acid than cyclohexanol. The pK_a values of some phenols are given below for comparison.

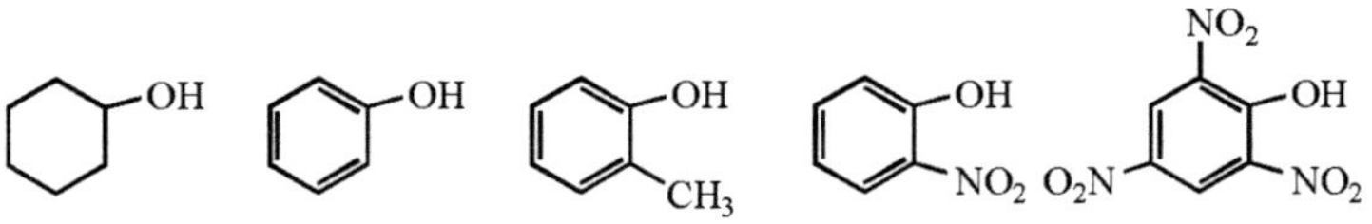

Cyclohexanol	Phenol	2-Methylphenol	2-Nitrophenol	2,4,6-Trinitrophenol
$pK_a = 18$	$pK_a = 9.9$	$pK_a = 10.2$	$pK_a = 7.2$	$pK_a = 0.38$

▶ *Key point* – The electron-releasing methyl group lowers the acid strength and the electron-withdrawing nitro group increases the acid strength. Thus, 2-methylphenol is a weaker acid than phenol, whereas 2-nitrophenol is a stronger acid than phenol. 2,4,6-Trinitrophenol (picric acid) is a fairly strong acid.

The greater acidity of a phenol with respect to an alcohol is due to the electron-withdrawing effect of the phenyl group. The phenoxide ion is more stable than alkoxide as a lone pair of electrons from oxygen is delocalized into the ring to form a resonance hybrid as follows:

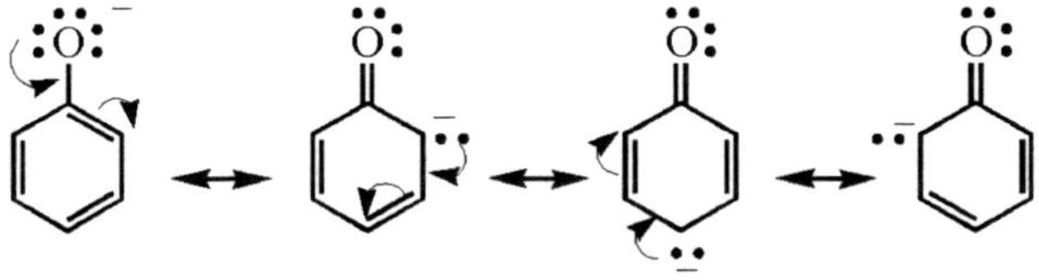

Reactions of phenols

1. Phenols form esters with acid chlorides and acid anhydrides in the presence of a base. Thus, phenol reacts with ethanoyl chloride (acetyl chloride) to give phenyl ethanoate (phenyl acetate).

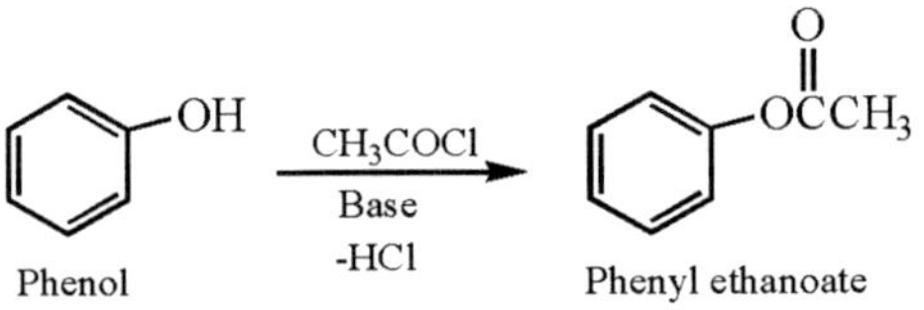

2. The hydroxyl group increases the reactivity of the phenyl group and thus phenol is more reactive than benzene. Phenol reacts with bromine water to give 2,4,6-tribromophenol and with dilute nitric acid to give a mixture of o- and p- nitrophenols.

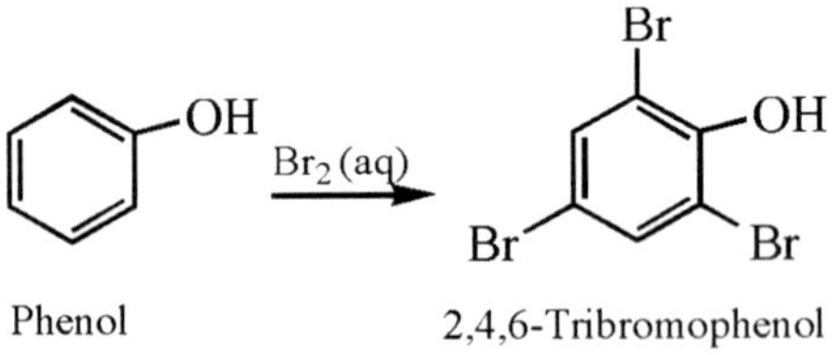

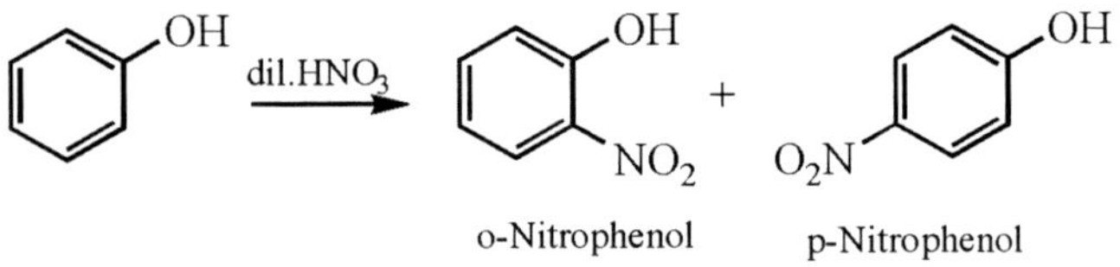

Tutorial: helping you learn

Progress questions

1. Draw the structural formulae of the constitutional isomers of the alcohols of molecular formula $C_5H_{12}O$, and give their IUPAC names. Group them under primary, secondary and tertiary alcohols. (Hint: There are eight isomeric alcohols of molecular formula $C_5H_{12}O$.)

2. Arrange the following compounds in the order of increasing pK_a values:

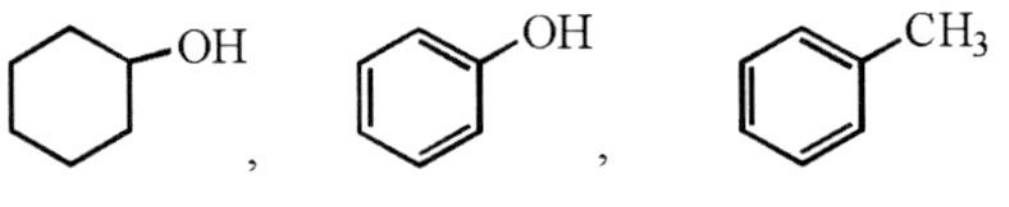

3. Arrange the following compounds in the order of increasing boiling points:

$CH_3CH_2CH_2CH_2CH_3$, $CH_3CH_2CH_2CH_2CH_2OH$, $CH_3CH_2CH_2CH_2CH_2Cl$

4. Write out the mechanism of the substitution reaction in each of the following cases:

   (a) The reaction of 1-butanol with HCl.
   (b) The reaction of 2,2-dimethyl-1-butanol with HBr.

5. Each of the following alcohols undergoes acid catalysed dehydration to give more than one product. Draw the structural formulae of the products in each case and indicate which of these is the major product. (Hint: The major product is the more stable alkene, i.e. the alkene which has the more substituted double bond.)

(a) 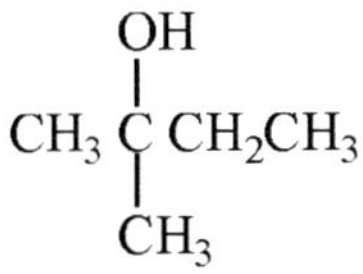

(b) 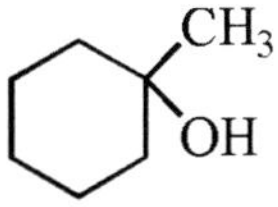

6. Suggest reagents and conditions for the following conversions:

(a) 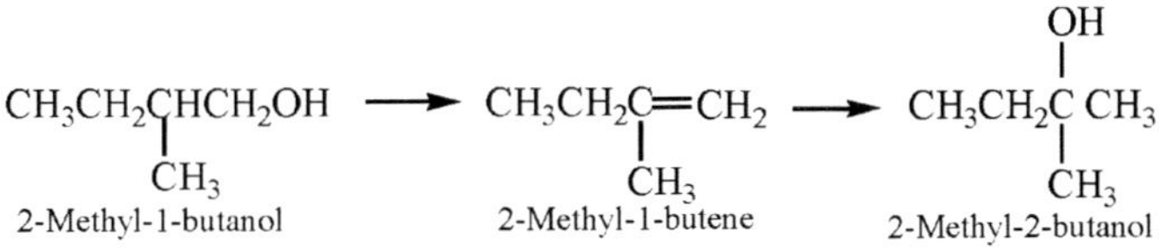

(b) 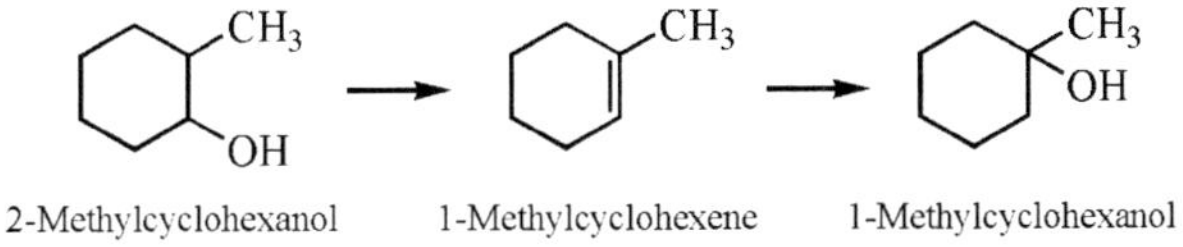

7. Phenol can be sulphonated with concentrated sulphuric acid.

   (a) Which has a greater affinity to sulphonation, benzene or phenol?
   (b) Is the hydroxyl group o-, p- directing or m- directing?
   (c) Draw the structures of the sulphonated products.

Discussion points

1. Alcohols can be dehydrated by treatment with acids, but phenols cannot be dehydrated under similar conditions. Account for this difference.

2. Alcohols can be converted into alkyl halides by treating with hydrogen halides, but phenols cannot be converted to the corresponding halogenocompounds with hydrogen halides. Account for this difference in terms of the strengths of carbon-oxygen bond. (Hint: The resonance structures of the phenoxide ion (Section 7.7) shows the double bond character between carbon and oxygen in phenoxide ion.)

Practical assignment

Make a comparative study of the properties and reactions of alcohols and phenols. Find the industrial uses of simple alcohols and phenols.

8

Aldehydes and Ketones

One-minute summary – Aldehydes and ketones are oxidation products of primary alcohols and secondary alcohols respectively. Both classes of compounds contain a carbonyl (C=O) group in common and thus share many properties. The carbonyl group is polar, having a δ-ve oxygen and a δ+ve carbon. Many nucleophiles attack the δ+ve carbon and electrophiles attack the δ-ve oxygen of the carbonyl group, resulting in the formation of various products. Methods of preparation of aldehydes and ketones, reactions with nucleophilic and electrophilic reagents and other reactions are discussed. Aldehydes are good reducing agents and are oxidised by a range of oxidising agents whereas ketones are resistant to oxidation. This difference in property is due to the presence of a hydrogen atom attached to the carbonyl carbon in aldehydes. In this chapter, we will cover:

- the carbonyl group
- methods of preparation of aldehydes and ketones
- nucleophilic additions at the carbonyl group
- reactions with amino compounds
- keto-enol tautomerism
- oxidation of aldehydes and ketones.

8.1 What are aldehydes and ketones?

Aldehydes are compounds containing a -CHO group and ketones, a -CO group (Refer Section 2.15). The general formula for aldehydes can be written as RCHO, and for ketones as R_2CO, where R stands for an alkyl or an aryl group.

Aldehydes are the oxidation products of primary alcohols, while ketones are the oxidation products of secondary alcohols.

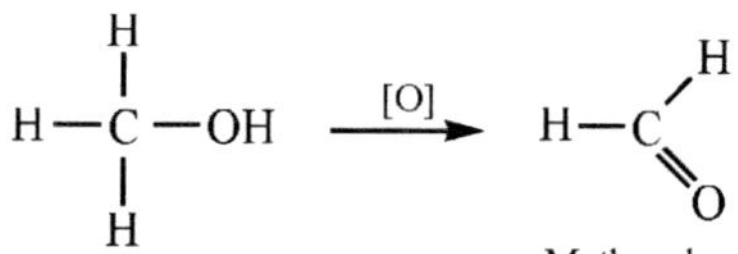

Methanol (Primary alcohol) → Methanal (Aldehyde)

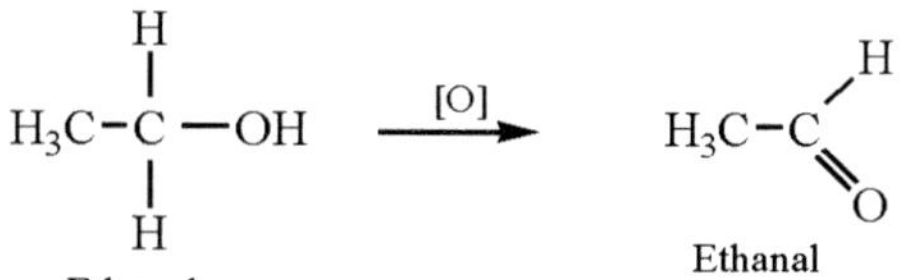

Ethanol (Primary alcohol) → Ethanal (Aldehyde)

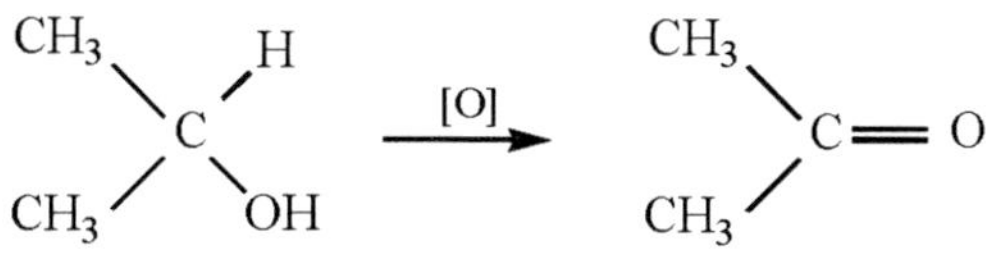

2-Propanol (Secondary alcohol) → Propanone (Ketone)

The simplest aldehyde is methanal (formaldehyde) and the simplest ketone is propanone (acetone). The IUPAC nomenclature of aldehydes and ketones is discussed in Section 2.15.

8.2 The structure of the carbonyl group

The carbonyl carbon is sp^2 hybridised, as in alkenes and the three sp^2 orbitals form three σ bonds, one with the carbonyl oxygen and the other two with hydrogen or carbon atoms. All three σ bonds are in one plane. The unhybridised *p* orbital of the carbon overlaps with a similar *p* orbital of oxygen to form a π bond. Thus, the carbonyl double bond is similar to the C=C double bond in alkenes. However, due to the polarity induced by the electronegative oxygen, it is more reactive to many reagents.

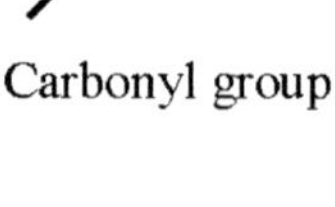

Carbonyl group

The carbon of the carbonyl group is therefore electrophilic and can react with nucleophiles. The oxygen is nucleophilic and can react with electrophiles.

The melting and boiling points of aldehydes and ketones are higher than those of hydrocarbons of similar molecular weights. This is due to the presence of the polar carbonyl group which allows fairly strong intermolecular **dipole-dipole interactions**. For example, the boiling point of butanone (molecular weight 72) is 80°C, while that of pentane (molecular weight 72) is 36°C.

▶ *Key point* – Dipole-dipole attraction is the attraction between the δ+ve end of a dipole and the δ-ve end of another dipole.

Lower molecular weight aldehydes and ketones are miscible with water. This is because the carbonyl oxygen can form hydrogen bonds with water molecules. Methanal, ethanal, propanal, propanone and butanone are soluble in water.

Aldehydes and ketones undergo many similar reactions due to their common carbonyl group. They differ in some properties because the aldehydic carbon is attached to a hydrogen atom. An aldehydic carbonyl group is less stabilised than a ketonic carbonyl group just as a primary carbocation is less stable than a secondary or tertiary carbocation. Aldehydes are thus more reactive than ketones. Another reason for the greater reactivity of aldehydes is that the aldehydic group is less hindered to approaching nucleophiles because it is bonded to only one alkyl group, while in ketones the carbonyl carbon is bonded to two bulky alkyl groups.

8.3 Preparation of aldehydes and ketones

1. By the oxidation of alcohols

As discussed in Section 7.6, chromium trioxide and potassium dichromate in aqueous sulphuric acid oxidise a primary alcohol to an aldehyde, and a secondary alcohol to a ketone. For low-molecular-weight aldehydes, the aldehyde formed is distilled from the reaction mixture as soon as it is produced, thus avoiding the oxidation of the aldehyde to a carboxylic acid.

$$\underset{\text{1-Butanol}}{CH_3CH_2CH_2CH_2OH} \xrightarrow{CrO_3} \underset{\text{Butanal}}{CH_3CH_2CH_2CHO}$$

$$\underset{\text{3-Pentanol}}{CH_3CH_2CHOHCH_2CH_3} \xrightarrow{CrO_3} \underset{\text{3-Pentanone}}{CH_3CH_2COCH_2CH_3}$$

There are some selective oxidising agents, one of which is pyridinium chlorochromate, $C_6H_5NH^+CrO_3Cl^-$, (PCC). When PCC in dichloromethane is used, the oxidation process stops at the aldehyde stage.

$$\underset{\text{1-Hexanol}}{CH_3\,(CH_2)_4\,CH_2OH} \xrightarrow[CH_2Cl_2]{PCC} \underset{\text{Hexanal}}{CH_3\,(CH_2)_4\,CHO}$$

2. By the ozonolysis of alkenes

An alkene which has a vinylic hydrogen (hydrogen attached to the double bonded carbon), on treatment with ozone, undergoes oxidative fission at the double bond to give an aldehyde.

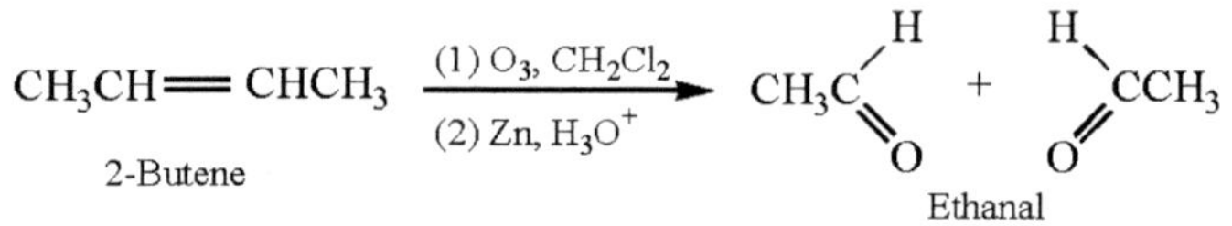

In the above reaction, since both carbon atoms at the double bond carry hydrogen atoms, two aldehyde molecules are formed. If the double bonded carbon has two alkyl substituents, a ketone is formed, as can be seen in the following example:

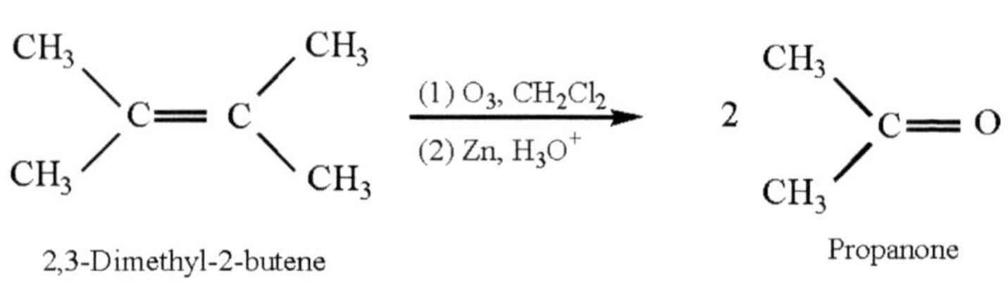

3. By Friedel-Crafts reaction

Aromatic ketones can be prepared by the Friedel-Crafts reaction. For example, when benzene is heated with acetyl chloride in the presence of an aluminium chloride catalyst, acetophenone is formed.

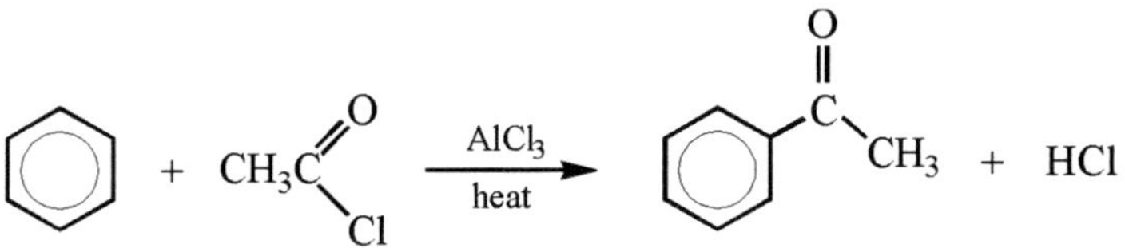

Benzene Acetyl chloride Acetophenone

4. By the hydration of alkynes

Alkynes add a molecule of water in the presence of mercuric sulphate in sulphuric acid to give ketones.

$$\underset{\text{2-Butyne}}{CH_3C\equiv CCH_3} + H_2O \xrightarrow[H_2SO_4]{HgSO_4} \underset{\text{Butanone}}{CH_3CH_2\overset{\displaystyle O}{\overset{\|}{C}}CH_3}$$

8.4 Reactions of aldehydes and ketones

In the following sections, some of the important reactions of aldehydes and ketones, including nucleophilic addition reactions, oxidation and reduction reactions, condensation reactions and Aldol reactions are explained.

8.5 Addition reactions

1. Addition of hydrogen (reduction)

When treated with suitable reducing agents, an aldehyde is reduced to a primary alcohol and a ketone to a secondary alcohol. In these reactions, the carbonyl group adds two hydrogen atoms. A number of reducing agents are available for the reduction reactions. These include hydrogen in the presence of nickel, platinum or palladium catalysts, sodium borohydride ($NaBH_4$) and lithium aluminium hydride ($LiAlH_4$, also called LAH).

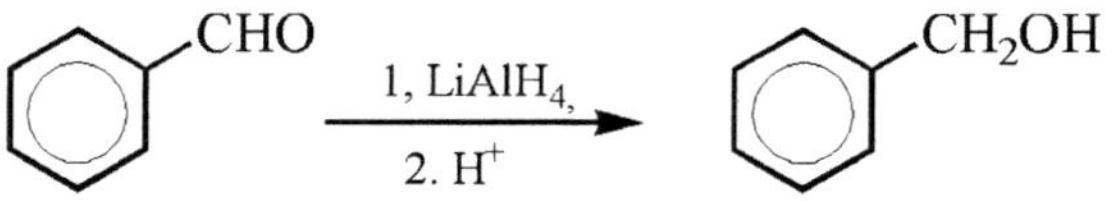

Benzaldehyde Benzyl alcohol (A primary alcohol)

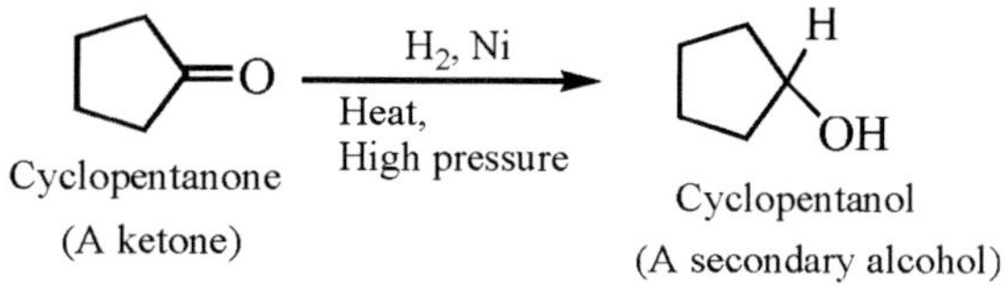

Sodium borohydride is used more widely than LAH as it is safer to handle and can be used in a reaction medium of water and alcohol. LAH reacts vigourously with water and alcohol; and thus solvents such as diethylether or tetrahydrofuran should be used.

The mechanism of a reduction reaction with borohydride is as follows: a hydride ion from the reducing agent is transferred to the positive carbonyl carbon; simultaneously an O-B bond is formed (with a pair of electrons at the double bond) to give an intermediate:

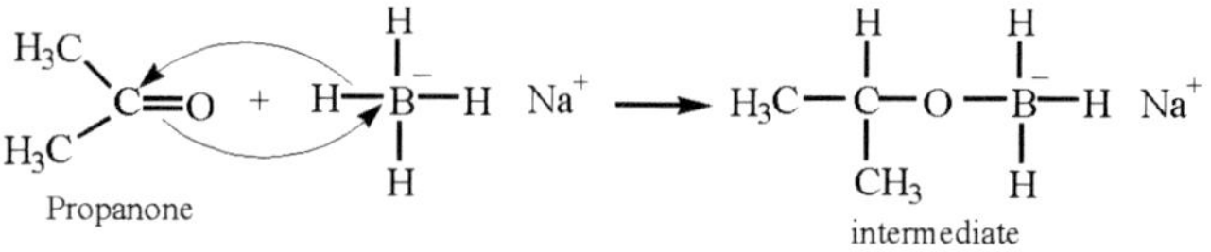

The intermediate hydrolyses to give an alcohol:

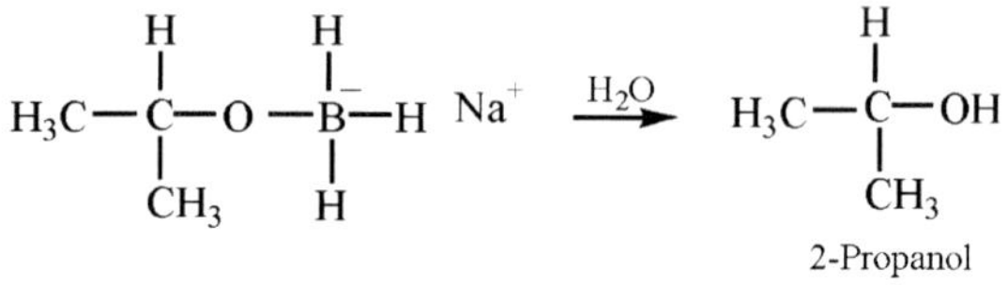

2. Addition of hydrogen cyanide

Aldehydes and ketones add hydrogen cyanide in the presence of sodium cyanide (which acts as a catalyst), to form addition products. A nucleophilic addition reaction takes place here. The mechanism of the reaction is explained in Section 4.6. The products formed are known by the general name **cyanohydrins**.

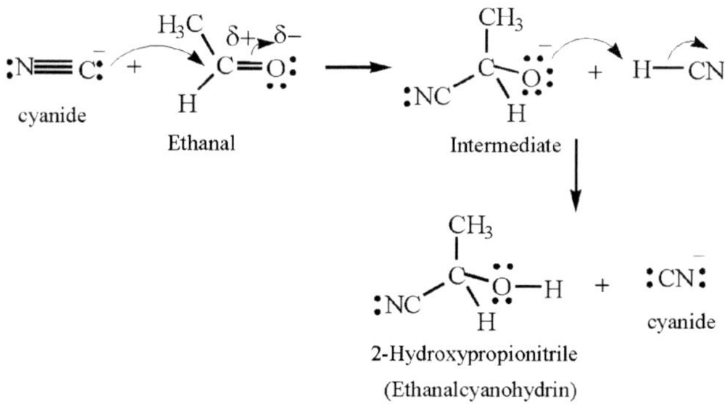

In summary, a molecule of H-CN is added in this reaction: H to the oxygen and CN to the carbon of the carbonyl group.

3. Addition of sodium hydrogen sulphite

Aldehydes and ketones react with sodium hydrogen sulphite ($NaHSO_3$) to form addition products. These usually precipitate out of the aqueous reaction mixture as crystalline solids.

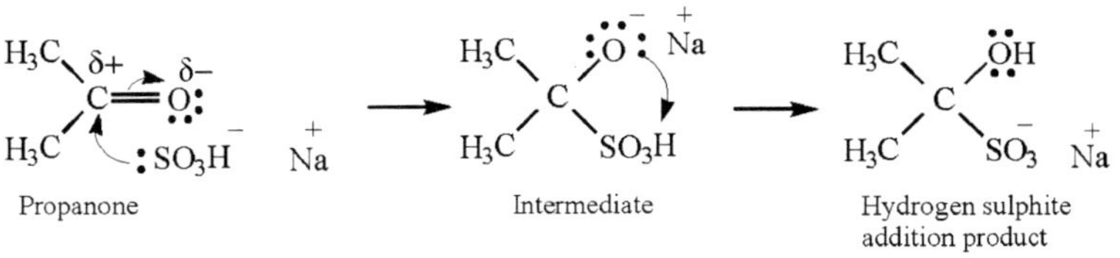

▶ *Key point* – Since the hydrogen sulphite addition products of aldehydes and ketones can be hydrolysed back to the aldehydes and ketones, this reaction is used for the separation or purification of aldehydes and ketones from mixtures of compounds containing them.

4. Addition of phosphoranes (the Wittig reaction)

Aldehydes and ketones react with phosphoranes to form alkenes. Phosphoranes, also called phosphorus ylides, are molecules which have a negative carbon adjacent to a positive phosphorus. An example of a phosphorane is shown below.

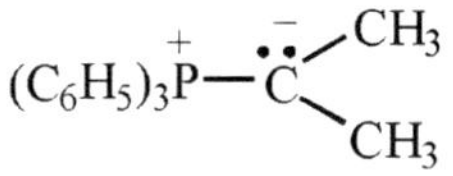

This particular phosphorane can be prepared by treating triphenylphosphine with 2-chloropropane in benzene, in the presence of a strong base such as phenyllithium.

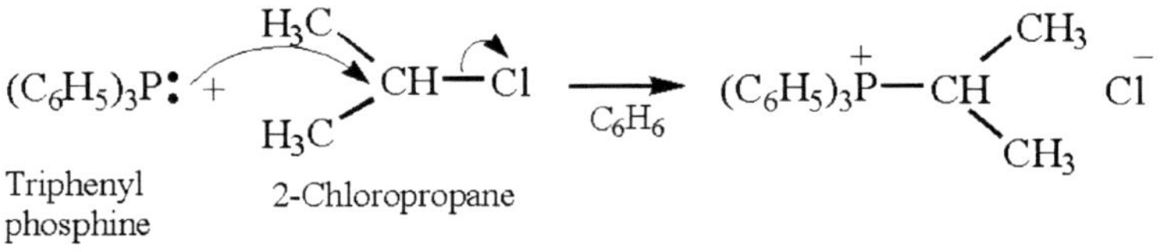

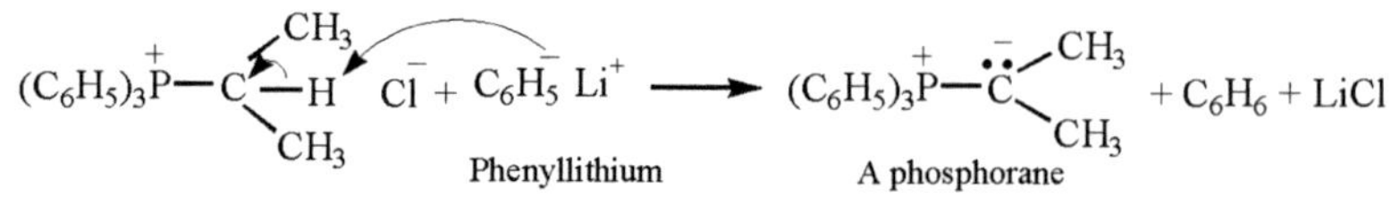

When this phosphorane reacts with acetophenone, the following reaction takes place to give an alkene:

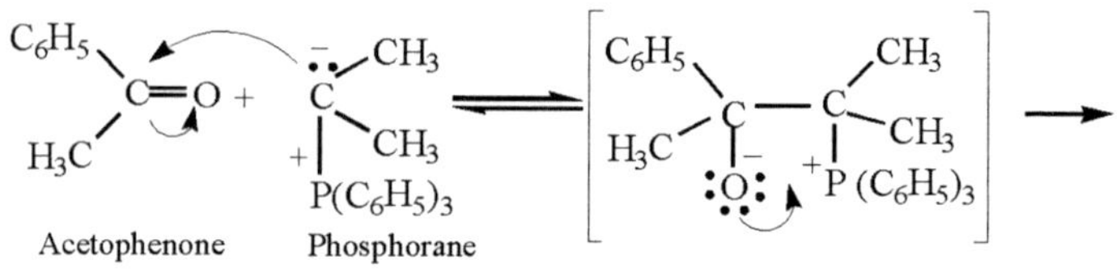

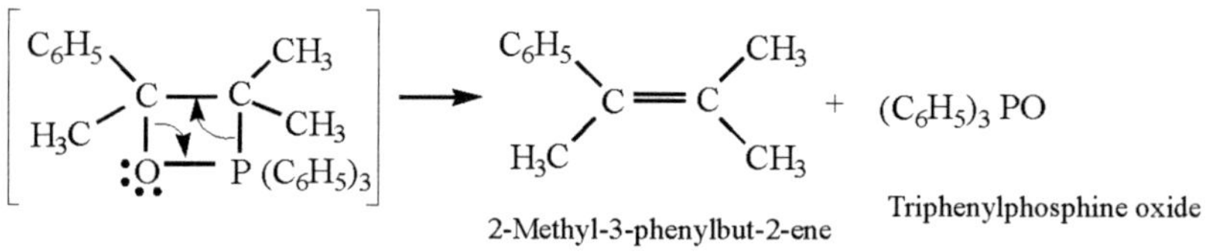

This reaction is very useful for the preparation of alkenes starting from aldehydes and ketones.

5. Addition of water

Aldehydes, especially the lower ones have a tendency to add water at the carbonyl group to form diols. This reaction is reversible. Ketones undergo hydration to a smaller extent.

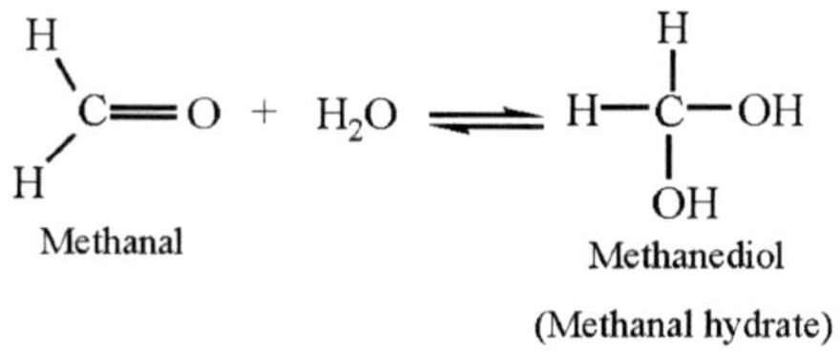

6. Addition of alcohols

Alcohols like water add to aldehydes in the presence of an acid to form compounds called **hemiacetals** which add another molecule of alcohol to give **acetals**. Under similar conditions, ketones give rise to **hemiketals** and **ketals**. Acetals and ketals contain two alkoxy (-OR) groups attached to the same carbon atom:

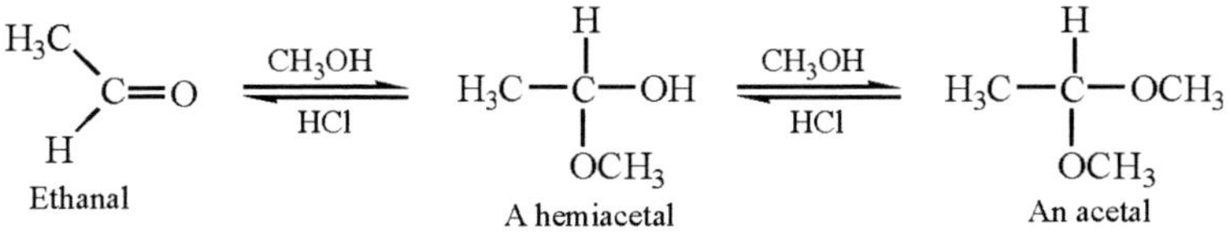

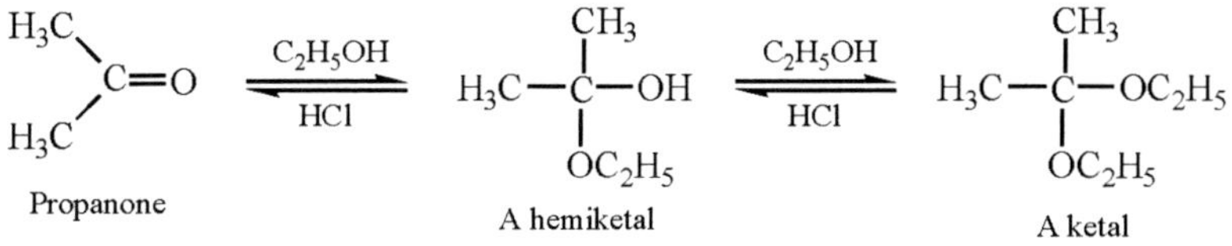

Acetals and ketals are stable in a basic medium, but hydrolysed in aqueous acid.

7. Addition of Grignard reagents

Aldehydes and ketones react with Grignard reagents by the nucleophilic addition reaction mechanism. Grignard reagents are alkyl or aryl magnesium halides of formula RMgX, where R can be an alkyl or aryl group and X, Cl, Br or I.

A Grignard reagent is prepared by heating an alkyl or aryl halide with magnesium in diethylether solvent.

$$C_6H_5Br + Mg \xrightarrow{(C_2H_5)_2O} C_6H_5MgBr$$

Bromobenzene → Phenylmagnesium bromide

The carbon-magnesium bond is highly polar as magnesium is highly electropositive and the carbon atom bonded to magnesium is negative. This carbon acts as a nucleophile and forms an addition product with the carbonyl compound, as shown below. The addition product hydrolyses with a dilute acid to give an alcohol. This method can be used to convert aldehydes and ketones to alcohols:

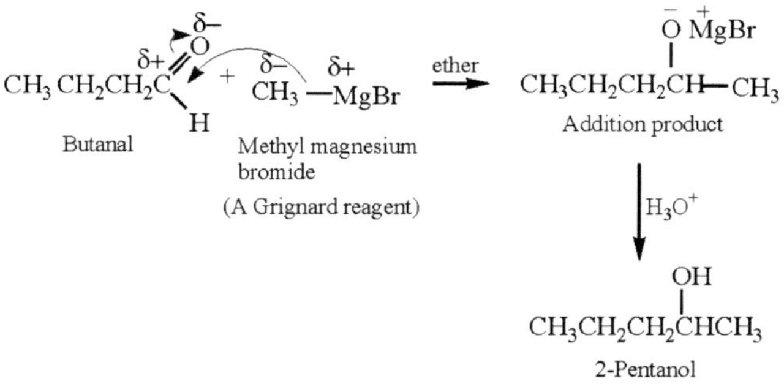

In the above reaction starting from butanal, a secondary alcohol 2-pentanol is prepared. Tertiary alcohols are similarly prepared from ketones and primary alcohols from methanal. Note these examples:

$$H_2C{=}O + C_2H_5MgBr \longrightarrow C_2H_5\,CH_2O^-\,MgBr^+ \longrightarrow C_2H_5CH_2OH$$

Methanal; Ethylmagnesium bromide; 1-Propanol

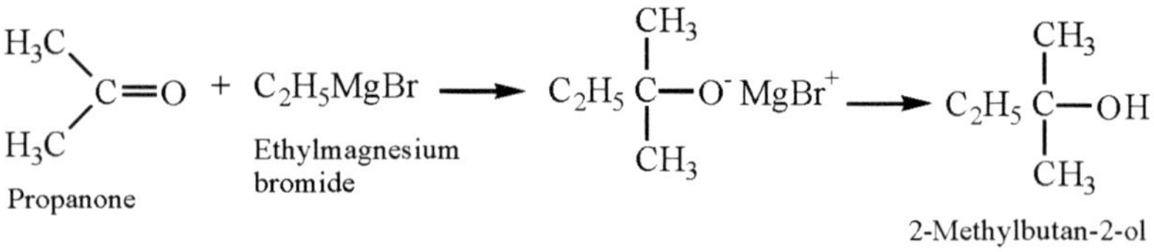

8.6 Condensation reactions

Aldehydes and ketones react with ammonia and a number of compounds which contain an amino (-NH_2) group to give condensation products. In these reactions, the nucleophilic amino group attacks the positive carbon of the carbonyl group to form an addition product which undergoes rearrangement followed by elimination of a molecule of water. So an addition/elimination reaction is said to take place.

The mechanism of these acid catalysed reactions is illustrated by taking the reaction between propanone and hydrazine.

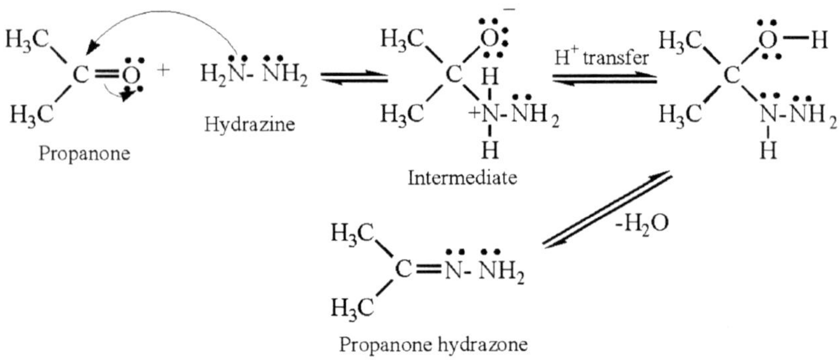

1. With hydroxylamine

Aldehydes and ketones react with hydroxylamine to give condensation products called oximes:

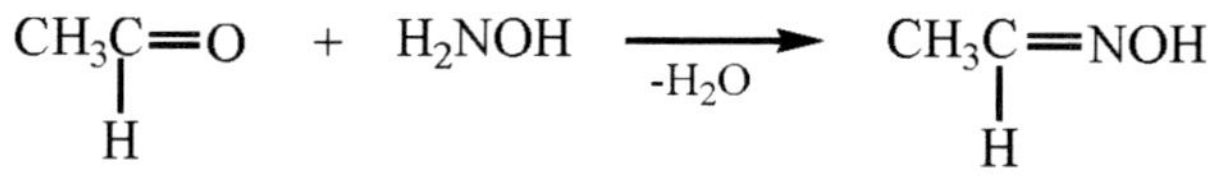

Ethanal Hydroxylamine Ethanaloxime

2. With hydrazines

Aldehydes and ketones react with hydrazine, phenyl hydrazine and 2,4-dinitrophenyl hydrazine to give hydrazones:

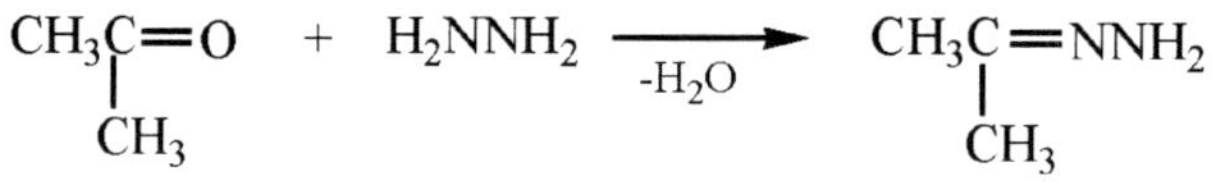

Propanone Hydrazine Propanonehydrazone

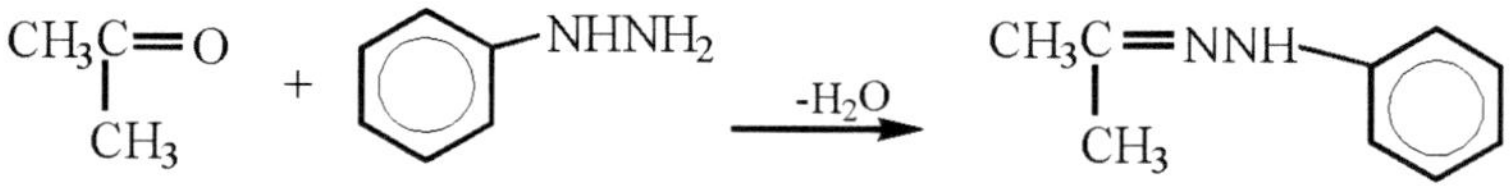

Propanone Phenylhydrazine Propanonephenylhydrazone

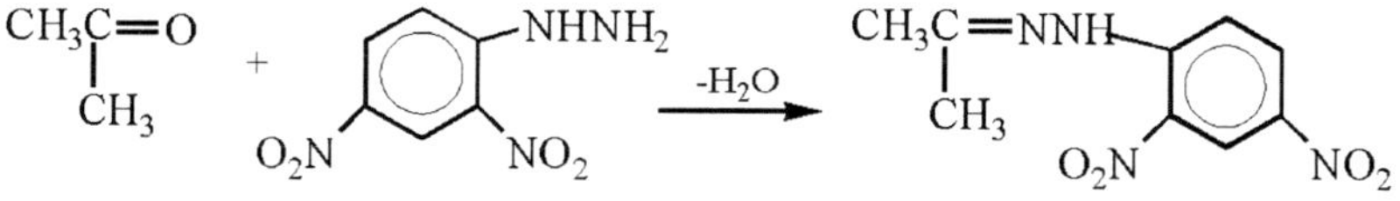

Propanone 2,4-Dinitro-phenylhydrazine Propanone-2,4-dinitro phenylhydrazone

► *Key point* – 2,4-Dinitrophenylhydrazine is also used as a laboratory reagent to test for the carbonyl group. 2,4-Dinitrophenylhydrazones are formed as yellow to orange coloured precipitates.

8.7 Keto-enol tautomerism

The α-hydrogen atoms of aldehydes and ketones (hydrogen atoms attached to the carbon adjacent to the carbonyl group) are acidic. One reason for the acidity of the α-hydrogen is that the electron-withdrawing inductive effect of the carbonyl group polarises the C-H bond. A strong base can abstract a proton from the polarised C-H bond. The resultant anion (**enolate**) stabilises due to the delocalisation of the negative charge, and the formation of a resonance hybrid of two structures, as shown below:

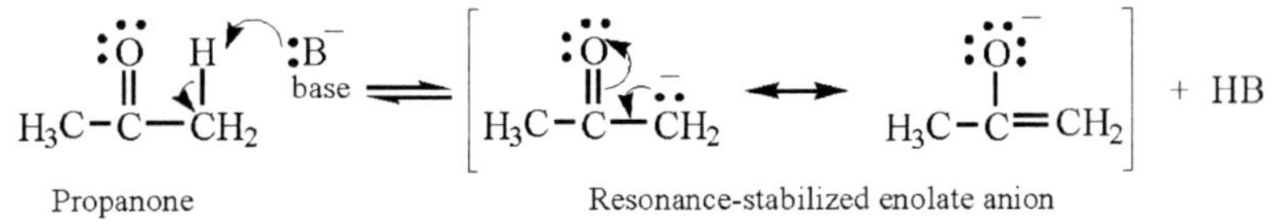

The anions in the resonance hybrid can accept protons to form the original carbonyl molecule or its enol form. (**Enols** are compounds containing an -OH attached to a double bonded carbon.)

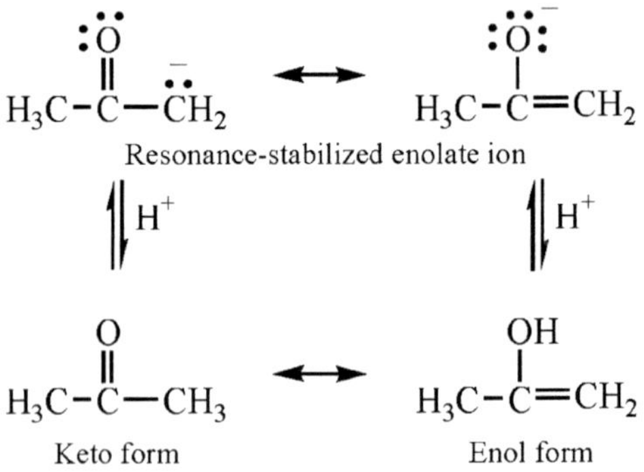

The **keto** and **enol** forms of a compound are examples of **tautomers**. **Tautomers** are rapidly interconvertible constitutional isomers which differ in the location of a hydrogen and a double bond. **Tautomerism** is the type of isomerism exhibited by keto-enol tautomers. Keto-enol tautomerism is catalysed by bases (as shown above) as well as by acids (as given below).

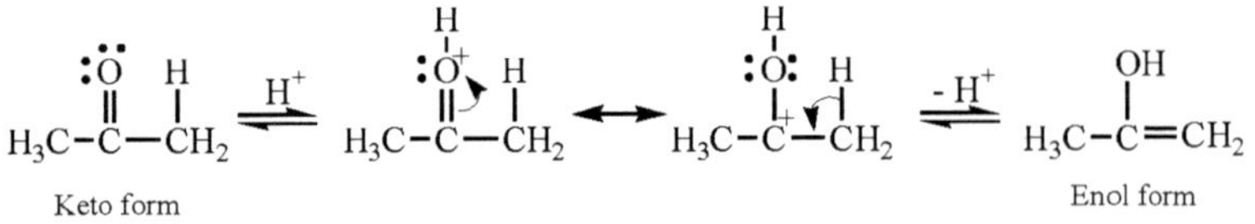

α-Halogenation of aldehydes and ketones

Aldehydes and ketones can be halogenated at the α position by treating with chlorine, bromine or iodine in an acid medium. When propanone is treated with bromine in ethanoic acid, acid-catalyzed enolization takes place first. The π electrons at the double bond polarise bromine, and a Br^+ ion adds to the enol using an electron pair from the enol. The resultant carbocation loses a proton from the -OH group to give the α-bromoketone.

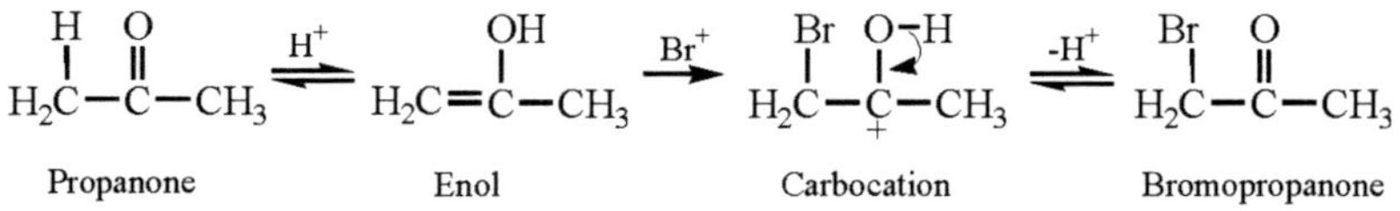

The Aldol reaction

We have learned that the α-hydrogen atom of an aldehyde or a ketone is acidic, and can be removed by a base to form an enolate anion. The enolate anion can attack a carbonyl carbon and undergo a nucleophilic addition reaction. When ethanal is treated with an ethanolic solution of sodium hydroxide, two molecules of ethanal condense to form **aldol** (3-hydroxybutanal). The reaction takes place through a number of steps. First, an α-hydrogen of ethanal is removed by the base to form the enolate ion. Then the enolate ion undergoes nucleophilic addition to a second molecule of ethanal, resulting in the formation of an intermediate which takes a proton from a proton donor to give the product.

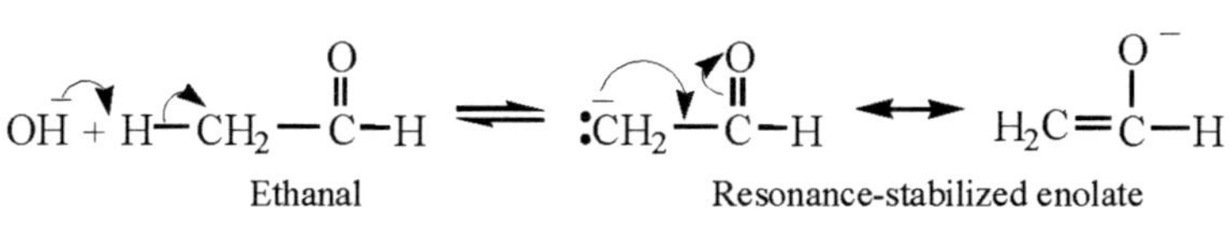

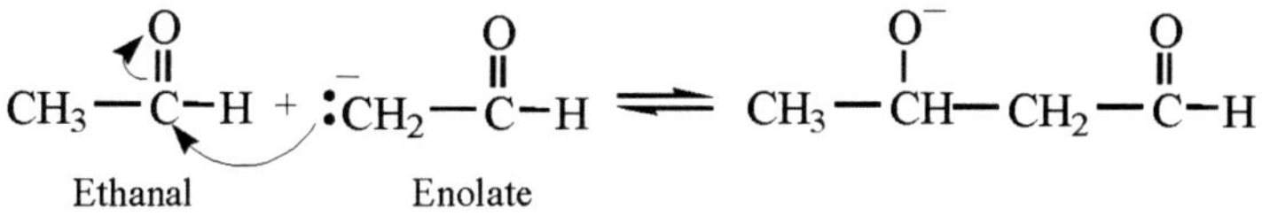

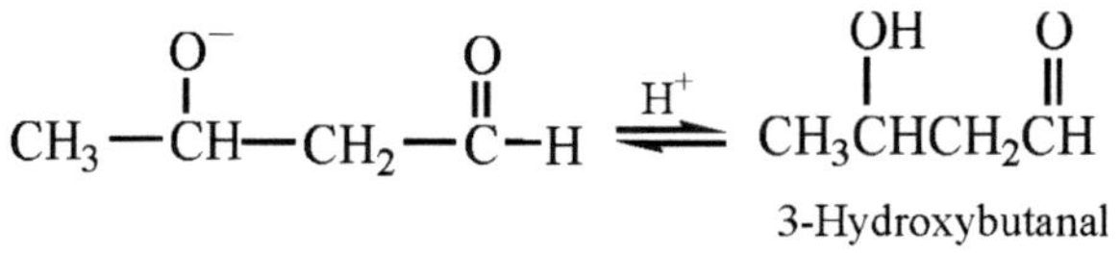

This type of addition reaction, in which two molecules of an aldehyde or a ketone combine to form a molecule of β-hydroxyaldehyde or ketone, is called an **aldol reaction**. An aldol reaction cannot take place without an α-hydrogen.

8.8 Oxidation of aldehydes and ketones

1. Aldehydes are good reducing agents and are easily oxidised by a number of oxidising agents to give the corresponding carboxylic acids.

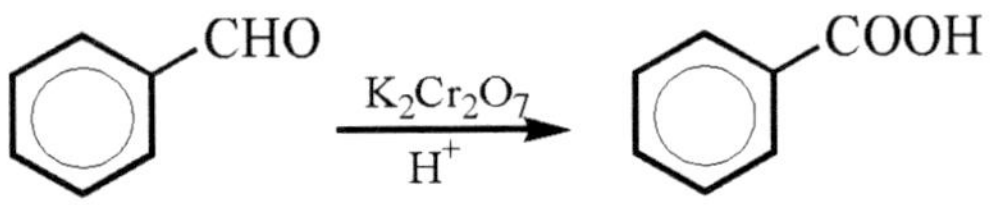

Benzaldehyde — Benzoic acid

2. Aldehydes reduce even mild oxidising agents like silver ion to silver. The reaction is carried out by mixing an aldehyde with an ammoniacal solution of silver nitrate. The aldehyde is oxidised to a carboxylate ion and the silver formed during the reaction is deposited on the side of the test tube as a shiny mirror. This test is thus called the **silver mirror test**. Ammoniacal silver nitrate solution is called Tollen's reagent and is prepared by adding sodium hydroxide solution to aqueous silver nitrate to form a precipitate, which is then dissolved by adding aqueous ammonia.

$$\underset{\text{Ethanal}}{CH_3CHO} + 2Ag(NH_3)_2^+ + H_2O \longrightarrow \underset{\text{Ethanoate}}{CH_3COO^-} + 2Ag + 3NH_4^+ + NH_3$$

3. Ketones are more resistant to oxidation. However, strong oxidising agents at higher temperatures can oxidise a ketone into a mixture of carboxylic acids, each containing fewer carbon atoms than the starting ketone. The double bond in the enol form of the molecule breaks to give the oxidation products.

$$\underset{\text{3-Pentanone}}{CH_3CH_2\overset{\displaystyle O}{\overset{\|}{C}}\!-\!CH_2CH_3} \rightleftharpoons \underset{\text{Enol form}}{CH_3CH_2\underset{\displaystyle OH}{\underset{|}{C}}\!=\!CHCH_3} \longrightarrow \underset{\text{Propanoic acid}}{CH_3CH_2CO_2H} + \underset{\text{Ethanoic acid}}{HO_2CCH_3}$$

4. Aldehydes and ketones are oxidised by peroxy acids. This is called **Baeyer-Villiger oxidation**. An example of this reaction is the oxi-

dation of acetophenone to phenyl acetate by treating it with peroxybenzoic acid.

$$\underset{\text{Acetophenone}}{C_6H_5\overset{\overset{O}{\|}}{C}CH_3} \xrightarrow{C_6H_5\overset{\overset{O}{\|}}{C}OOH} \underset{\text{Phenyl acetate}}{C_6H_5{-}O{-}\overset{\overset{O}{\|}}{C}CH_3}$$

Tutorial: helping you learn

Progress questions

1. Draw the formula of an isomeric ketone in each of the following cases and give their IUPAC names.

(a) $CH_3CH_2CH_2CHO$

Butanal

(b) 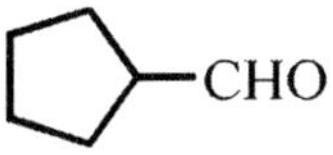

Cyclopentanecarbaldehyde

(c) $CH_3\overset{\overset{CH_3}{|}}{C}H\,\underset{\underset{CH_3}{|}}{C}H\,COCH_3$

3,4-Dimethyl-2-pentanone

(d) 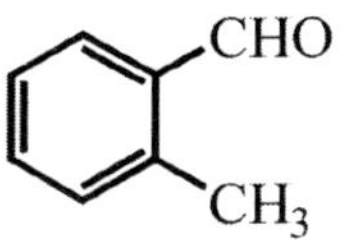

2-Methylbenzaldehyde

2. Write equations for the reactions between the following compounds:

   (a) Propanal and 2,4-dinitrophenylhydrazine.
   (b) Hexanal and hydrogen cyanide.
   (c) 1-Phenyl-2-propanol and pyridinium chlorochromate in dichloromethane.
   (d) Cyclopentanone and phenyl magnesium bromide (C_6H_5MgBr), followed by hydrochloric acid.

3. Propose a method to prepare each of the following compounds:

   (a) 2-Hydroxybutanenitrile ($CH_3CH_2CHOHCN$).
   (b) Hexanoic acid from an aldehyde.
   (c) 2-Pentanone from an alkyne.
   (d) 1-Phenylethanol from a ketone.

4. Explain the mechanism for the reactions between the following substances:

   (a) 3-Pentanone and methanol in the presence of traces of HCl to form a hemiacetal.
   (b) Cyclohexanone and vinyl magnesium bromide ($CH_2{=}CHMgBr$).

5. Identify products A to E.

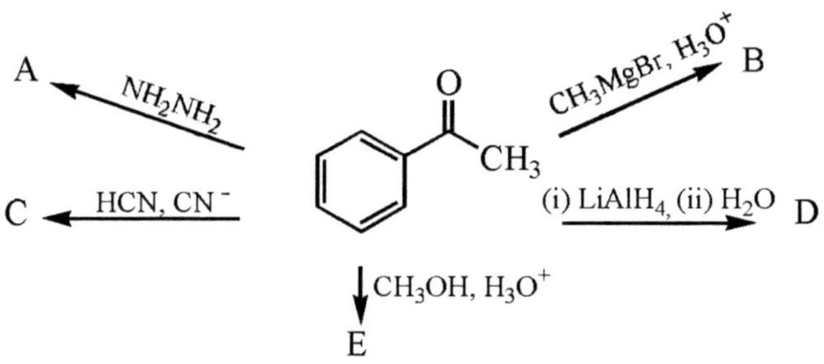

Discussion point

1-Phenylprop-1-ene ($C_6H_5CH{=}CHCH_3$) can be prepared by the Wittig reaction using two different phosphoranes. Outline the synthesis of this compound from these phosphoranes.

Practical assignment

1. Methanal (formaldehyde) is used as a disinfectant and as a preservative. It is also used in the industrial preparations of polyurethane polymers and antioxidants. Investigate the starting materials and the reactions in these syntheses.

2. By now, you must be familiar with the writing of the reaction mechanisms. Take the reactions (the reaction mechanisms of which are not given) in this chapter, and practice writing the reaction mechanisms.

9

Carboxylic Acids and their Derivatives

One-minute summary – The functional group in carboxylic acids is the **carboxyl** (-COOH) group which consists of a carbonyl group and a hydroxyl group. The two electron withdrawing oxygen atoms attached to the carbon make it positive, and thus susceptible to nucleophilic attack. So carboxylic acids undergo nucleophilic substitution reactions leading to the formation of acid chlorides, esters, acid anhydrides and amides. Though alcohols contain hydroxyl groups, they are very weak acids, but carboxylic acids are moderately strong. Studying the structure of the carboxyl group will help you understand the nature of carboxylic acids. In this chapter, we will cover:

- the structure and nature of the carboxyl group
- methods of preparation of carboxylic acids
- reactions of carboxylic acids
- reduction of the carboxyl group
- acid derivatives.

9.1 Carboxylic acids

Carboxylic acids are compounds which contain a -COOH group. The -COOH group consists of a carbonyl group as in aldehydes and ketones and a hydroxyl group as in alcohols. The general formula of aliphatic carboxylic acids is RCOOH (or RCO_2H) where R stands for an alkyl group; and of aromatic acids, $ArCO_2H$, where Ar stands for an aryl group.

Nomenclature

The IUPAC nomenclature of carboxylic acids is discussed in Section 2.16. Many carboxylic acids are commonly known by their trivial names. Some important acids and their IUPAC and common names are as follows:

Trivial name	*IUPAC name*	*Structure*
Formic acid	Methanoic acid	HCO_2H
Acetic acid	Ethanoic acid	CH_3CO_2H
Propionic acid	Propanoic acid	$CH_3CH_2CO_2H$
Oxalic acid	Ethanedioic acid	HO_2CCO_2H
Malonic acid	Propane-1,3-dioic acid	$HO_2CCH_2CO_2H$

Table 9.1. Some carboxylic acids.

9.2 Structure and properties

One of the two oxygen atoms in the carboxyl group (-COOH) is bonded to the carbon by a double bond and the other oxygen is in the hydroxyl group bonded to the carbon and a hydrogen atom.

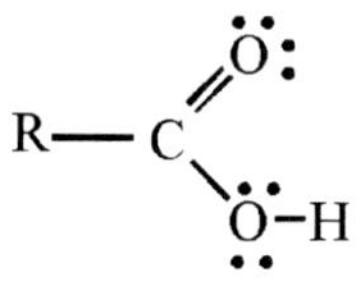

Structure of carboxylic acid group

The carbon atom of the carboxyl group is sp^2 hybridised. So the carboxyl group is planar, and the C-C-O and O-C-O bond angles are nearly 120° as would be expected for an sp^2 hybridised carbon. The -COOH group is polarised with δ+ve carbon, δ+ve hydrogen and δ-ve oxygen atoms.

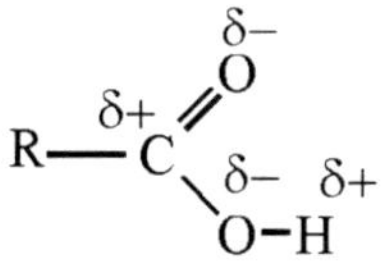

The molecules form intermolecular hydrogen bonds through the negative oxygen and positive hydrogen. This accounts for the high melting and boiling points of carboxylic acids.

Carboxylic acid molecules dimerise through hydrogen bonding as shown below, and in fact exist in this form in liquids and solids.

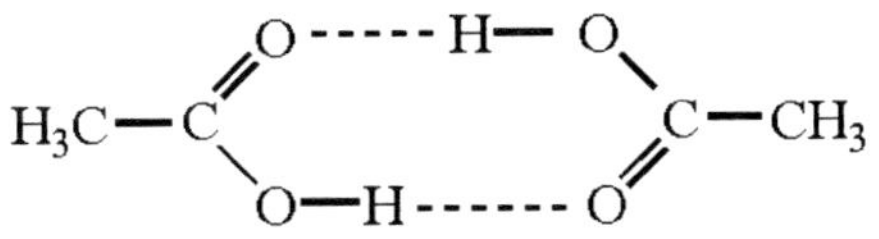

Dimerized ethanoic acid molecules

The lower members of the carboxylic acids are miscible with water, again due to the hydrogen bonds developed between the carboxylic acid and water. Solubility decreases as the length of the carbon chain increases.

9.3 Acidity of carboxylic acids

Carboxylic acids dissociate in water according to the equation,

$$RCO_2H + H_2O \rightleftharpoons RCO_2^- + H_3O^+$$

The acid strength depends on the extent of the forward reaction and is measured in terms of an acid constant, K_a.

$$K_a = \frac{[RCO_2^-]\,[H_3O^+]}{[RCO_2H]}$$

The higher the K_a value, the stronger the acid. As discussed earlier in Section 7.2, pK_a values are used to compare acid strengths. The higher the pK_a value, the weaker the acid.

A carboxylic acid dissociates to form a carboxylate ion (RCO_2^-) which is a resonance-stabilised anion, the negative charge being concentrated over the two oxygen atoms. The ion may be represented by the two contributing forms shown below.

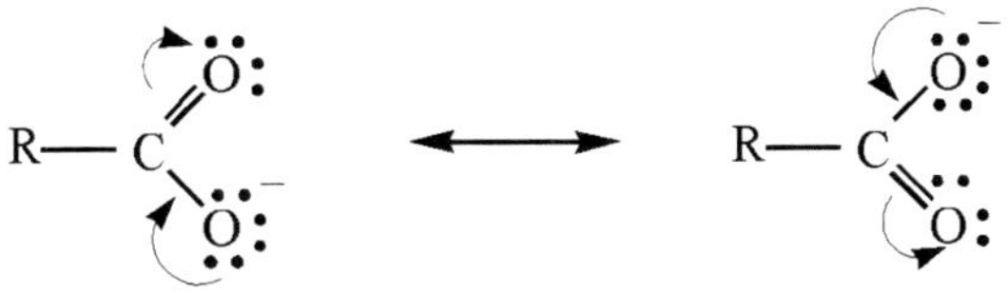

The structures which contribute to the resonance hybrid of the carboxylate ion

Carboxylic acids are stronger acids than alcohols though they both contain -OH group. One reason for the difference in acid strength is the presence of a second oxygen atom bonded to the carboxyl carbon which draws electrons towards it, thus making the O-H bond weak. Another reason is that the carboxylate ion is resonance stabilised, thus favouring the dissociation of the acid, which is not possible with the alkoxide ion.

Substituents increase or decrease the acidity of the carboxylic acids. The electron donating inductive effect of alkyl groups slightly destabilises the carbanion, makes the O-H bond less polar and thus reduces the extent of dissociation, lowering the acid strength. In the same way, electron withdrawing inductive effect of groups such as halogens, nitro, vinyl and phenyl weakens the O-H bond and increases the acid strength. The halogen-substituted ethanoic acids are stronger than ethanoic acid. As the electronegativity of the halogen increases, the acid strength increases. As the number of halogen atoms increases, the acid strength increases.

Acid	*Structure*	pK_a
Methanoic	HCO_2H	3.75
Ethanoic	CH_3CO_2H	4.72
Fluoroethanoic	CH_2FCO_2H	2.59
Chloroethanoic	CH_2ClCO_2H	2.85
Bromoethanoic	CH_2BrCO_2H	2.90
Iodoethanoic	CH_2ICO_2H	3.18
Dichloroethanoic	$CHCl_2CO_2H$	1.26
Trichloroethanoic	CCl_3CO_2H	0.85
Benzoic	$C_6H_5CO_2H$	4.19
4-Nitrobenzoic	$(4\text{-}NO_2)C_6H_4CO_2H$	3.41

Table 9.2. Some carboxylic acid pK_a values.

9.4 Preparation of carboxylic acids

1. By the oxidation of primary alcohols and aldehydes

We have discussed the oxidation of primary alcohols (Section 7.6) and that of aldehydes (Section 8.7) in earlier chapters. Two examples are given below.

$$\underset{\text{1-Hexanol}}{CH_3(CH_2)_4CH_2OH} \xrightarrow[H_2SO_4]{K_2Cr_2O_7,} \underset{\text{Hexanoic acid}}{CH_3(CH_2)_4CO_2H}$$

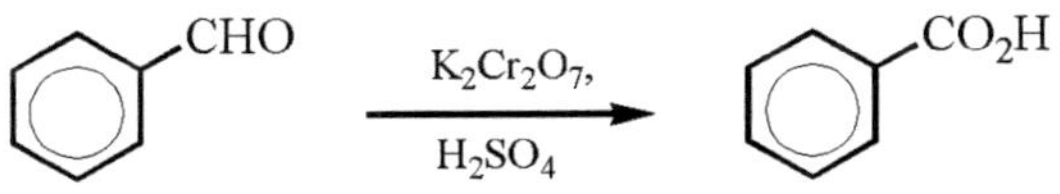

Benzaldehyde — Benzoic acid

2. By the side chain oxidation of aromatic compounds

Carboxylic acids can be produced by the oxidation of alkyl substituted benzenes. Primary and secondary alkyl groups directly bonded to the benzene ring undergo oxidation.

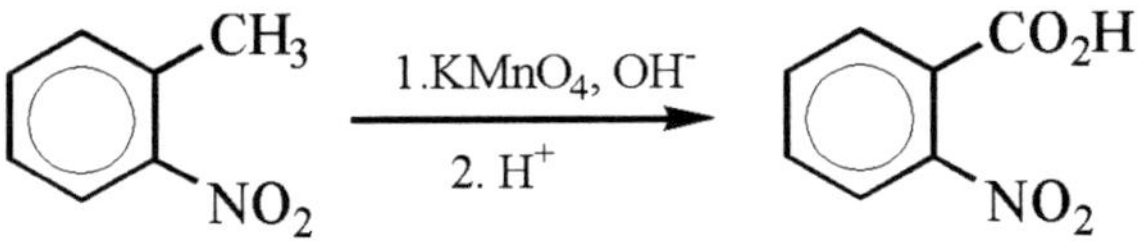

2-Nitrotoluene — 2-Nitrobenzoic acid

3. By the reaction of Grignard reagents with CO_2

A Grignard reagent adds a molecule of carbon dioxide and the addition product hydrolyses with dilute acids to give a carboxylic acid. For example, the Grignard reagent prepared by heating 4-bromotoluene with magnesium in diethylether, gives 4-methylbenzoic acid, according to the following scheme:

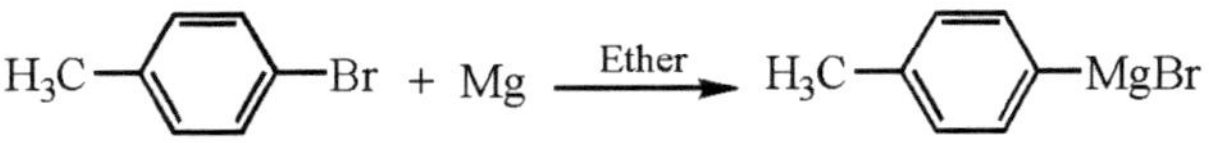

4-Bromotoluene — 4-Methylphenylmagnesium bromide

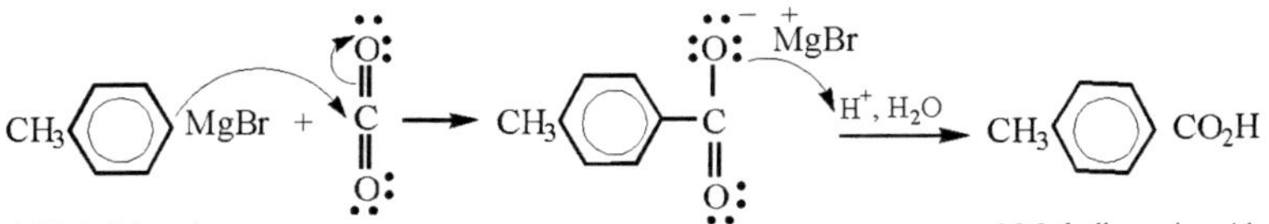

4-Methylphenylmagnesium bromide (A Grignard reagent) — Addition product — 4-Methylbenzoic acid

4. By the hydrolysis of nitriles

Nitriles, on acid or alkaline hydrolysis yield carboxylic acids. Nitriles are often prepared from halogenocompounds by treating with sodium cyanide.

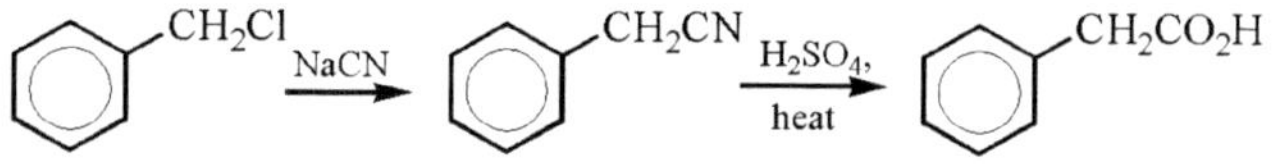

Benzyl chloride — Benzyl cyanide — Phenyl acetic acid

5. By the oxidation of alkenes

Alkenes on oxidation with alkaline potassium permanganate give a mixture of carboxylic acids.

$$\underset{\text{3-Heptene}}{CH_3CH_2CH_2CH{=}CHCH_2CH_3} \xrightarrow[2.\ H_3O^+]{1.\ KMnO_4,\ OH^-,\ heat} \underset{\text{Butanoic acid}}{CH_3CH_2CH_2CO_2H} + \underset{\text{Propanoic acid}}{HO_2CCH_2CH_3}$$

9.5 Reactions of carboxylic acids

1. Salt formation

Carboxylic acids react with bases to form salts.

$$\underset{\text{Ethanoic acid}}{CH_3CO_2H} + NaOH \longrightarrow \underset{\text{Sodium ethanoate}}{CH_3CO_2^- Na^+} + H_2O$$

Benzoic acid — Sodium benzoate

Carboxylic acids react with aqueous sodium carbonate and sodium hydrogen carbonate to form salts, carbon dioxide and water.

$$\underset{\text{Ethanoic acid}}{CH_3CO_2H} + NaHCO_3 \longrightarrow \underset{\text{Sodium ethanoate}}{CH_3CO_2^- Na^+} + H_2O + CO_2\,(g)$$

▶ *Key point* – The reaction between a carboxylic acid and sodium hydrogen carbonate takes place with effervescence and the evolution of carbon dioxide. Since carbon dioxide can be easily identified, this reaction is used to test for carboxylic acids. Note that alcohols and most phenols are not acidic enough to react with sodium hydrogen carbonate.

2. Conversion to acid chlorides and bromides

Carboxylic acids react with thionyl chloride ($SOCl_2$), phosphorus pentachloride (PCl_5) and phosphorus trichloride (PCl_3) to form acid chlorides. Heating carboxylic acids with a mixture of phosphorus and bromine gives rise to acid bromides. These are known by the general name *acyl halides*, and are formed by replacing the hydroxyl group of the acid by a halogen.

$$3\,CH_3C(=O){-}OH + PCl_3 \xrightarrow{\text{heat}} 3\,CH_3C(=O){-}Cl + H_3PO_3$$

Ethanoic acid (Acetic acid) → Ethanoyl chloride (Acetyl chloride)

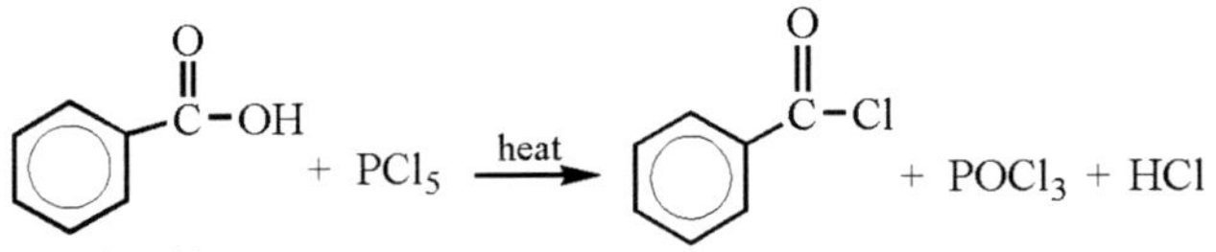

Benzoic acid → Benzoyl chloride

$$CH_3(CH_2)_4C(=O){-}OH + SOCl_2 \xrightarrow{\text{reflux}} CH_3(CH_2)_4C(=O){-}Cl + SO_2 + HCl$$

Hexanoic acid → Hexanoyl chloride

3. Formation of esters

Carboxylic acids react with alcohols in the presence of mineral acids to form esters. For example, when ethanoic acid is heated with methanol in the presence of concentrated sulphuric acid, methyl ethanoate is formed.

$$CH_3C(=O){-}OH + CH_3OH \underset{\text{heat}}{\overset{H_2SO_4}{\rightleftharpoons}} CH_3C(=O){-}O{-}CH_3 + H_2O$$

Ethanoic acid, Methanol → Methyl ethanoate

This reaction is reversible and the forward reaction is favoured by the use of sulphuric acid which removes the molecule of water formed during the reaction.

The mechanism of the reaction is illustrated below. In this reaction, the carboxylic acid undergoes protonation and this makes the carboxyl carbon more positive. This then bonds with the alcoholic oxygen.

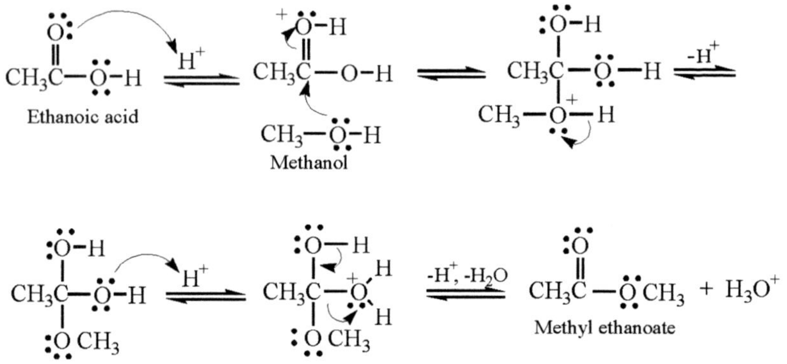

Hints and tips

1. The mechanism illustrated above shows that methanol forms a bond with the carboxyl carbon by a nucleophilic attack.
2. It is the hydroxyl group of the carboxylic acid that is removed from the intermediate to form water, and not the hydroxyl group of the alcohol.

4. Formation of amides

Carboxylic acids react with ammonia and amines to give ammonium salts which on heating decompose to give amides.

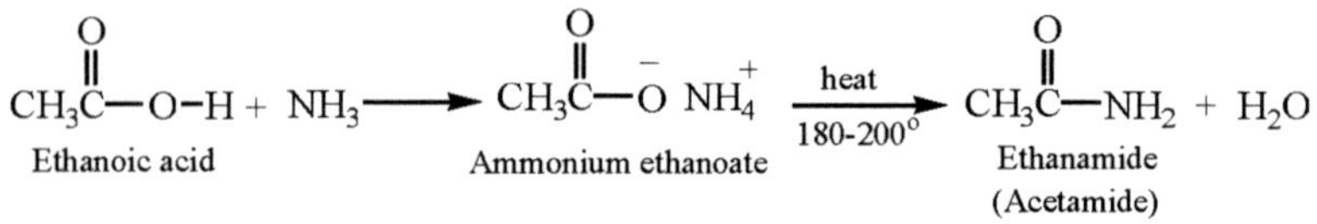

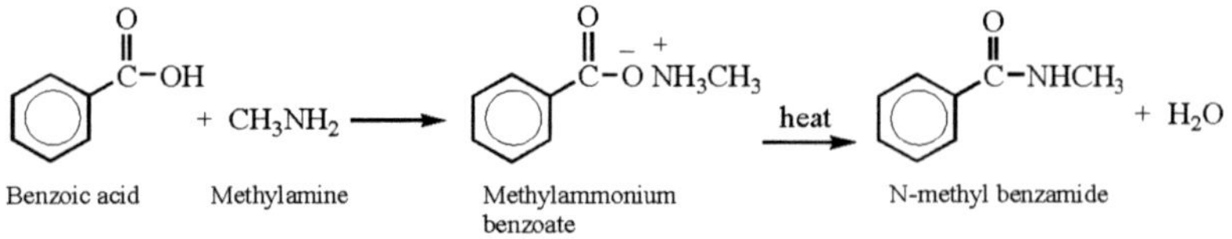

5. Reduction

Carboxylic acids are reduced to primary alcohols using lithium aluminium hydride ($LiAlH_4$) in diethylether. (Refer Section 8.5)

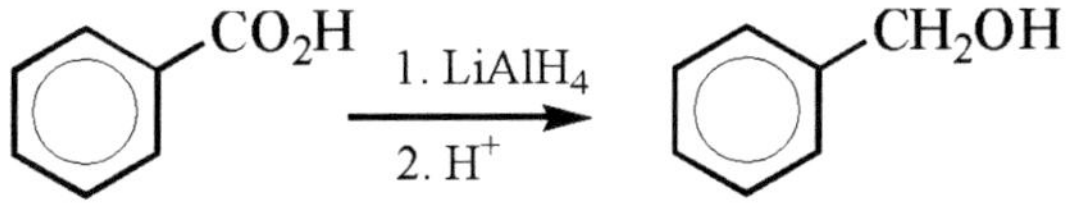

Benzoic acid　　　　Benzyl alcohol

6. α-Halogenation of aliphatic carboxylic acids

One or more hydrogen atoms of the α-carbon of aliphatic carboxylic acids can be substituted by chlorine or bromine when treated in the presence of a small quantity of phosphorus which acts as a catalyst.

$$\underset{\text{Ethanoic acid}}{CH_3CO_2H} + Cl_2 \xrightarrow[2.\ H_2O]{1.\ P} \underset{\text{Chloroethanoic acid}}{CH_2ClCO_2H}$$

7. Decarboxylation

Carboxylic acids can be decarboxylated by heating with soda lime which is a mixture of sodium hydroxide and calcium oxide. The strong base removes a molecule of CO_2 to give the corresponding decarboxylated product.

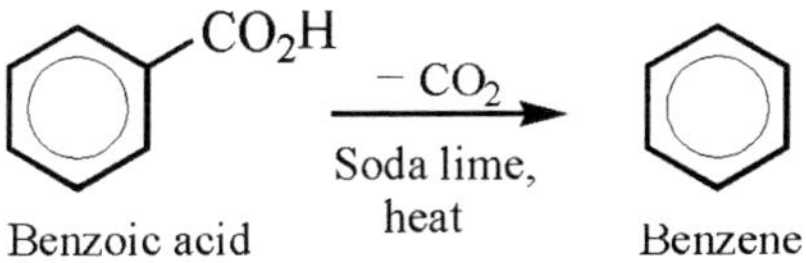

Benzoic acid　　　　Benzene

9.6 Derivatives of carboxylic acids

Acid halides or acyl halides

Acid halides are compounds which contain a -COX group (X = Cl, Br or I). Acid halides are formed by substituting the hydroxyl group of a carboxylic acid by a halogen. The preparation of acid halides is discussed in Section 9.5.

Acid halides are named from the corresponding carboxylic acids by replacing the *-ic acid* ending by *-yl halide.* Note the names of the following acid halides.

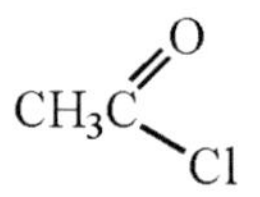

Ethanoyl chloride (Acetyl chloride)

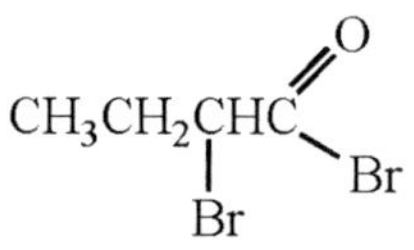

2-Bromobutanoyl bromide

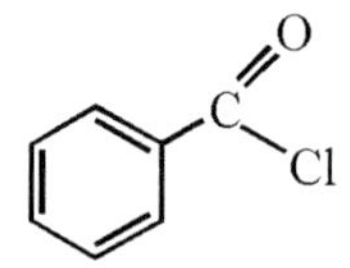

Benzoyl chloride

Acid chlorides are very reactive. They are hydrolysed easily to give carboxylic acids:

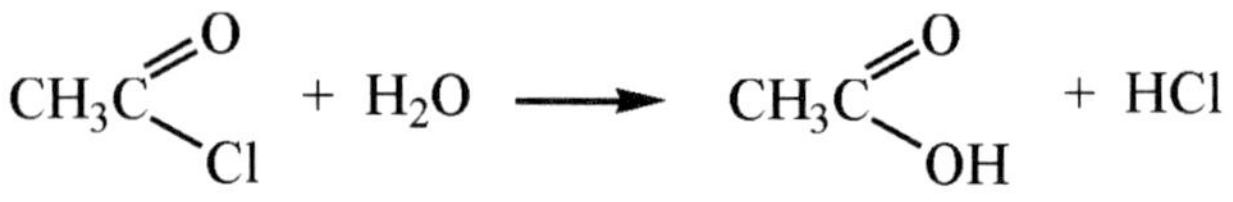

Ethanoyl chloride — Ethanoic acid

Acid halides are used in the preparation of acid anhydrides, esters and amides. Examples are given in the following sections.

Carboxylic acid anhydrides

Carboxylic acid anhydrides are condensation products of two acid molecules formed by the elimination of a molecule of water.

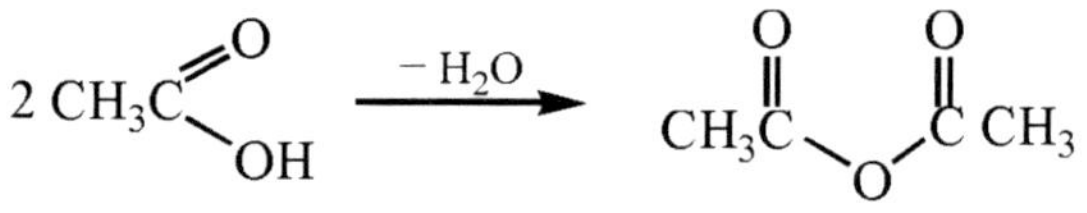

Ethanoic acid (Acetic acid) — Ethanoic anhydride (Acetic anhydride)

Acid anhydrides can be prepared more easily by treating carboxylic acids with acid chlorides, than by dehydrating carboxylic acids. The reaction is carried out in the presence of a base like pyridine, C_5H_5N, the function of which is to remove HCl formed during the reaction.

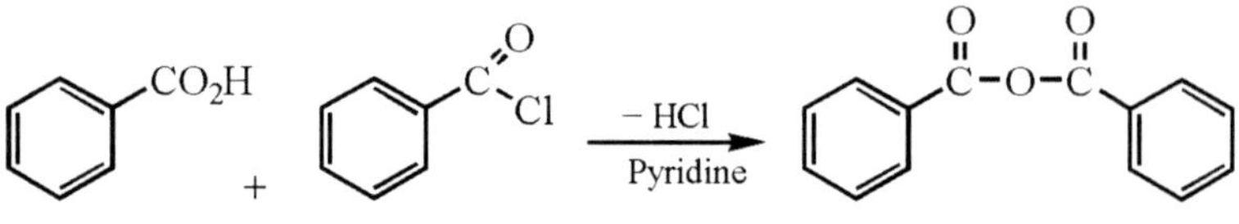

Benzoic acid — Benzoyl chloride — Benzoic anhydride

Acid anhydrides are hydrolysed back into two molecules of carboxylic acid. Acid anhydrides are also very reactive, but not as much as acid halides.

Esters

Esters are compounds which contain a -COOR group, i.e. carbonyl carbon attached to an -OR group. They are prepared by the reaction of carboxylic acids with alcohols in the presence of a catalyst (Section 9.5), or by treating an acid halide with an alcohol in the presence of a base. The second method is commonly used for the preparation of esters, since acid halides are very reactive and can be prepared easily.

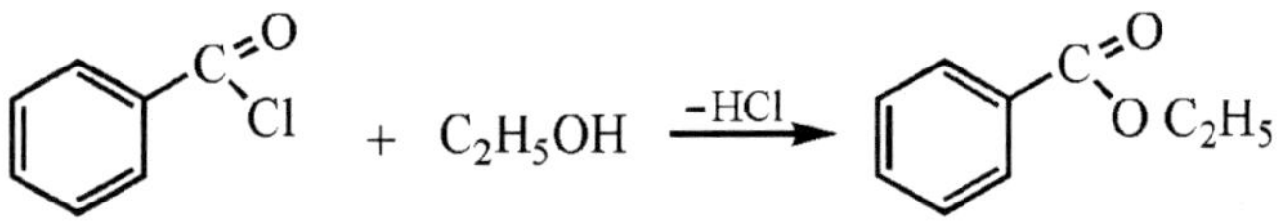

Benzoyl chloride Ethanol Ethyl benzoate

Esters undergo acid hydrolysis to give a carboxylic acid and an alcohol.

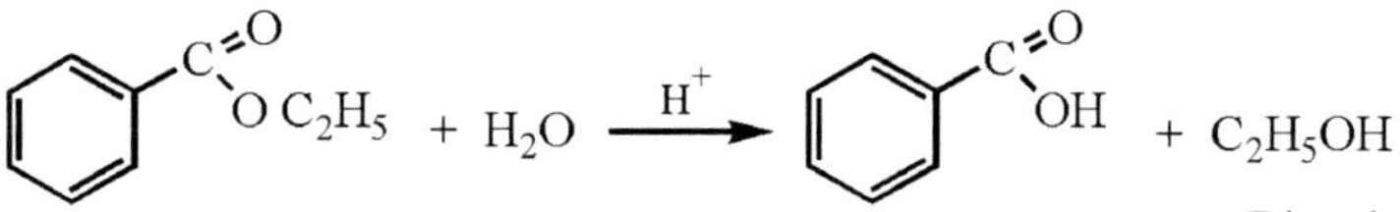

Ethyl benzoate Benzoic acid Ethanol

The alkaline hydrolysis of esters gives rise to salts of carboxylic acids and alcohols. This reaction is called **saponification** (soap formation, since soaps are sodium or potassium salts of higher carboxylic acids, and are prepared by the alkaline hydrolysis of esters of higher carboxylic acids).

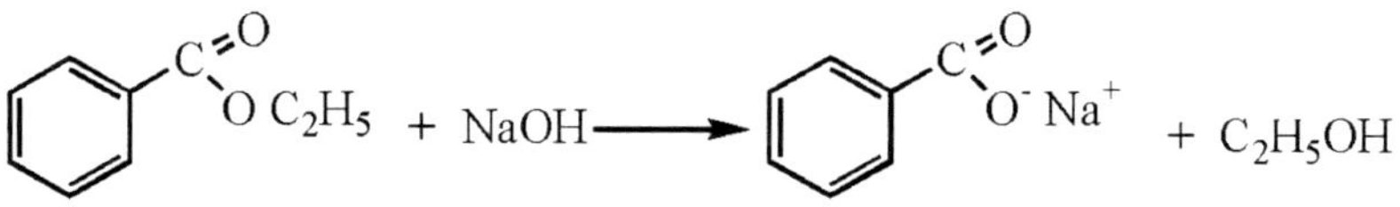

Ethyl benzoate Sodium benzoate Ethanol

Amides

Amides are condensation products of the reaction of carboxylic acids with amines, and contain a $-CONH_2$ group in primary amides, a $-CONHR$ group in secondary amides, and a $-CONR_2$ group in tertiary amides. Below are some examples of amides:

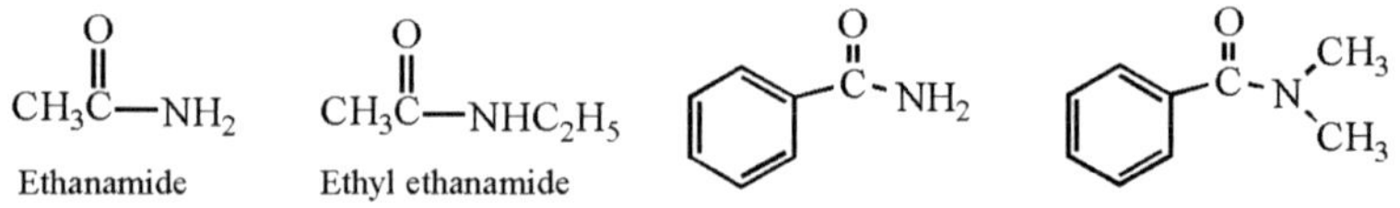

Ethanamide (Acetamide)	Ethyl ethanamide (Ethyl acetamide)	Benzamide	N,N-dimethylbenzamide
A primary amide	A secondary amide	A primary amide	A tertiary amide

Amides are prepared by heating ammonium carboxylates or alkylammonium carboxylates at about 200°C (Section 9.5), or by the reaction of acid chlorides with ammonia or amines. Excess ammonia or amine is used to neutralise the hydrogen halide.

$$\underset{\text{Ethanoyl chloride}}{CH_3\overset{O}{\overset{\|}{C}}Cl} + NH_3 \xrightarrow[NH_3]{-HCl} \underset{\text{Ethanamide}}{CH_3\overset{O}{\overset{\|}{C}}NH_2}$$

Amides are hydrolysed in the presence of acids or bases to carboxylic acids.

$$\underset{\text{Ethanamide}}{CH_3\overset{O}{\overset{\|}{C}}NH_2} + H_2O + HCl \longrightarrow \underset{\text{Ethanoic acid}}{CH_3\overset{O}{\overset{\|}{C}}\text{-}OH} + NH_4Cl$$

Tutorial: helping you learn

Progress questions

1. Draw the structural formula for each of the following compounds:

   (a) 2,2-Dichloropropanoic acid
   (b) 3-Nitrobenzoic acid
   (c) 2,4-Dimethyl-2-bromopentanoic acid
   (d) Phenyl ethanoic acid.

2. Give the reaction conditions and equations for the following conversions (some conversions take more than one step):

(a) $CH_3CH_2CH_2CH_2Cl \longrightarrow CH_3CH_2CH_2CH_2COOH$

1-Chlorobutane → Pentanoic acid

(b) 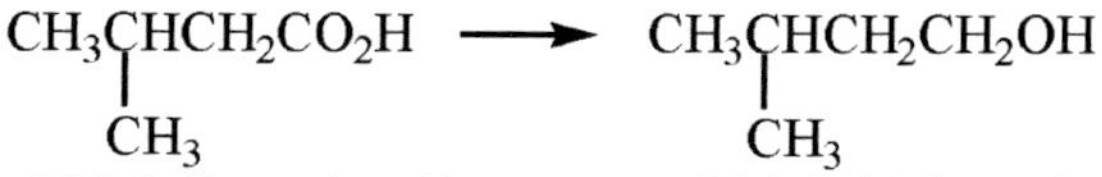

3-Methylbutanoic acid → 3-Methyl-1-butanol

(c) 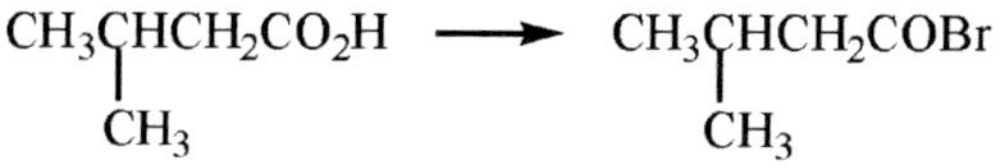

3-Methylbutanoic acid → 3-Methylbutanoyl bromide

(d) 

Ethylbenzene → Benzoic acid

2. Arrange the following compounds in the order of increasing acid strength:

(a) CH_3CO_2H, CH_2ClCO_2H, CH_2FCO_2H, CH_2BrCO_2H,
(b) CH_3CO_2H, $C_6H_5CO_2H$, p-$NO_2C_6H_4CO_2H$, C_6H_5OH

3. Acid chlorides are easily hydrolysed to carboxylic acids. Propose a reaction mechanism for the hydrolysis of ethanoyl chloride (CH_3COCl) to ethanoic acid.

4. Give the reaction products A–G:

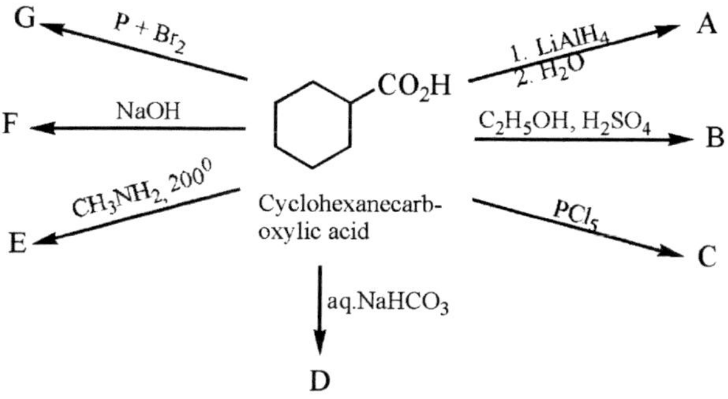

Discussion points

In the formation of the four main types of acid derivatives (acid halides, esters, acid anhydrides and amides), the hydroxyl group of the carboxylic acid is substituted. The mechanism of these reactions includes nucleophilic attack at the carboxyl carbon. Write the reaction mechanisms for the formation of the acid derivatives, taking appropriate examples. The mechanism of an ester formation reaction is given in Section 9.5.

Practical assignment

Waxes and fats are naturally occurring esters of long-chained fatty acids. Soaps are sodium or potassium salts of long-chained fatty acids. Look into the chemistry of soap formation from fats. Find the similarities and the differences between the structures and the cleaning action of soaps and detergents.

10

Amines: Organic Bases

One-minute summary – Amines are compounds derived by substituting one or more hydrogen atoms of ammonia with alkyl or aryl groups. They are bases and take a proton from water or an acid via the lone pair of electrons on the nitrogen atom. They are nucleophiles and react with alkyl halides and acyl halides. The structure and properties of amines and a comparative study of their basic strength are given. The objectives of this chapter are to discuss the:

- basic nature of amines
- salt formation reactions of amines
- reactions of amines with alkyl halides
- formation of amides from acid chlorides and anhydrides
- formation and synthetic importance of diazonium compounds.

10.1 Introducing amines

As mentioned above, amines are derivatives of ammonia. One or more hydrogen atoms are substituted by alkyl or aryl groups to form amines. Amines can be classified as primary, secondary and tertiary. A **primary amine** has the general formula RNH_2, and has one alkyl or aryl group attached to the -NH_2 group. A **secondary amine** has the general formula R_2NH, and a **tertiary amine**, R_3N (Section 2.17).

CH_3NH_2	$CH_3CH_2-N(CH_3)-H$	$CH_3-N(CH_3)-CH_3$
Methylamine (a primary amine)	Ethylmethylamine (a secondary amine)	Trimethylamine (a tertiary amine)

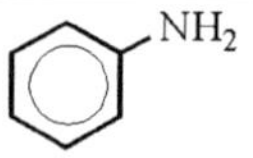

Phenylamine or Aniline
(a primary amine)

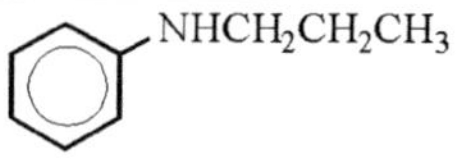

Phenylpropylamine or
N-propylaniline
(a secondary amine)

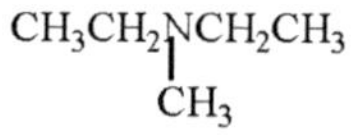

Diethylmethylamine
(a tertiary amine)

10.2 Properties of amines

Amine molecules are polar and can form intermolecular hydrogen bonds through the δ-ve nitrogen and δ+ve hydrogen of the amino group. As expected, amines have higher melting and boiling points than nonpolar compounds of similar molecular weight, but lower melting and boiling points than alcohols and carboxylic acids.

Amines are bases and they can accept a proton from water or an acid. The expression for the dissociation constant for a base, the **base constant**, K_b, is given below. For the dissociation of methylamine, CH_3NH_2, in water:

$$\underset{\text{Methylamine}}{CH_3NH_2} + H_2O \rightleftharpoons CH_3NH_3^+ + OH^-$$

$$K_b = \frac{[CH_3NH_3^+]\,[OH^-]}{[CH_3NH_2]}$$

$[H_2O]$ is taken as constant (Section 7.2). The higher the K_b value, the stronger the base. The pK_b value, which is the negative logarithm of K_b, is more often used for comparing base strengths. The higher the pK_b value, the weaker the base. Note the pK_b values of ammonia and some amines:

NH_3 4.75	CH_3NH_2 3.36	$CH_3CH_2NH_2$ 3.25	$CH_3CH_2CH_2NH_2$ 3.33
	CH_3NH_2 3.36	$(CH_3)_2NH$ 3.27	$(CH_3)_3N$ 4.19

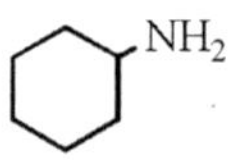

Cyclohexylamine
3.36

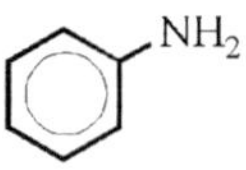

Aniline
9.42

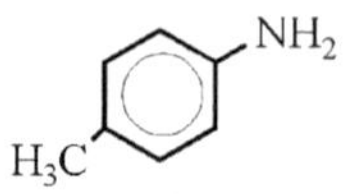

4-Methylaniline
8.92

pK_b values of some bases

It can be seen that aliphatic primary and secondary amines are stronger bases than ammonia while tertiary and aromatic amines are weaker. There are two main reasons for this trend in basic character:

1. Alkyl groups have an electron releasing inductive effect which makes the nitrogen of the amino group more negative, and thus increases its ability to protonate. A primary amine is therefore a stronger base than ammonia, and a secondary amine (with two alkyl groups attached to the nitrogen) is more basic than a primary amine. Aromatic amines are weaker bases because the lone pair of electrons on nitrogen are delocalised with the π electrons of the aromatic ring to form a resonance stabilised system (as shown below) and thus less available for protonation.

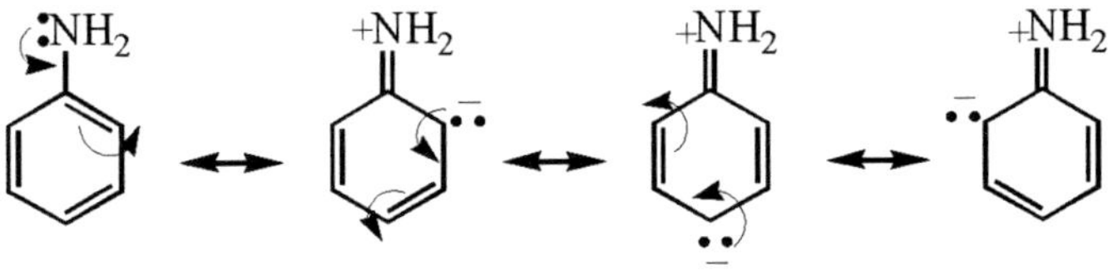

2. The second reason is the stabilisation of the ammonium ion formed through solvation or hydration. When an amine is protonated, an alkyl ammonium ion is formed. A primary ammonium ion (for example, $CH_3NH_3^+$) which has three hydrogen atoms bonded to the nitrogen can form three hydrogen bonds with water molecules and is thus more hydrated than an ion formed from a secondary amine. This factor favours the protonation of the base in water. As the number of hydrogen atoms bonded to the nitrogen atom of the ammonium ion decreases, the basic strength decreases. In the case of tertiary amines, though there are three alkyl groups, a lesser extent of solvation makes them weaker bases.

10.3 Methods of preparation of amines

1. By the reaction of alkyl halides with ammonia

When an alkyl halide is treated with excess ammonia, a primary amine is formed. For example, bromoethane reacts with ammonia to form ethylammonium bromide. A second molecule of ammonia removes a molecule of hydrogen bromide from ethylammonium bromide to give ethylamine:

$$CH_3CH_2Br + NH_3 \longrightarrow CH_3CH_2NH_3^+\,Br^- \xrightarrow{NH_3} CH_3CH_2NH_2 + NH_4Br$$

Bromoethane — Ethylamine

An alkyl halide reacts with a primary amine in the presence of ammonia to form a secondary amine, and with a secondary amine to give rise to a tertiary amine:

$$CH_3CH_2NH_2 + CH_3Br \xrightarrow{NH_3} CH_3CH_2NHCH_3 + NH_4Br$$

Ethylamine, Bromomethane — Ethylmethylamine

$$CH_3CH_2NHCH_3 + CH_3Br \xrightarrow{NH_3} CH_3CH_2N(CH_3)CH_3 + NH_4Br$$

Ethylmethylamine — Dimethylethylamine

2. By the reduction of nitrocompounds

Nitrocompounds can be reduced to amines by a number of reducing agents. Commonly used reducing agents are iron or tin with hydrochloric acid and hydrogen, in the presence of palladium or platinum catalyst.

$$CH_3CH_2CH_2NO_2 \xrightarrow{Fe,\ HCl} CH_3CH_2CH_2NH_2$$

1-Nitropropane — Propylamine

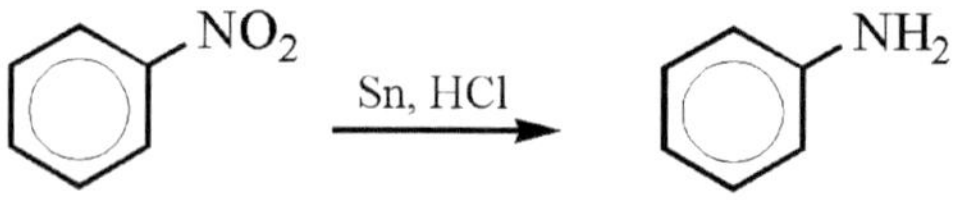

Nitrobenzene — Aniline

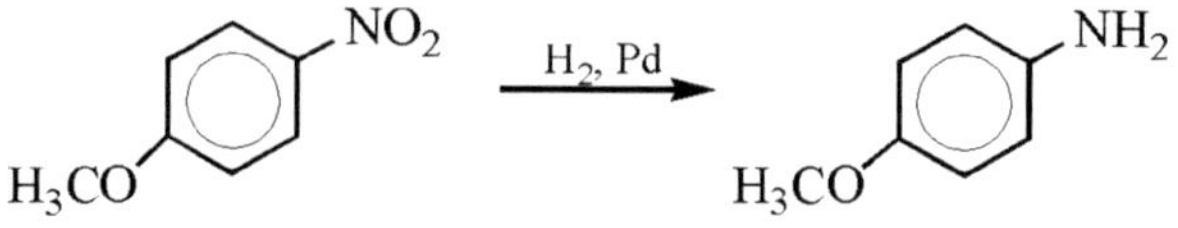

4-Methoxynitrobenzene — 4-Methoxyaniline

3. By the reduction of nitriles

Nitriles are reduced when treated with hydrogen in the presence of a catalyst to give primary amines:

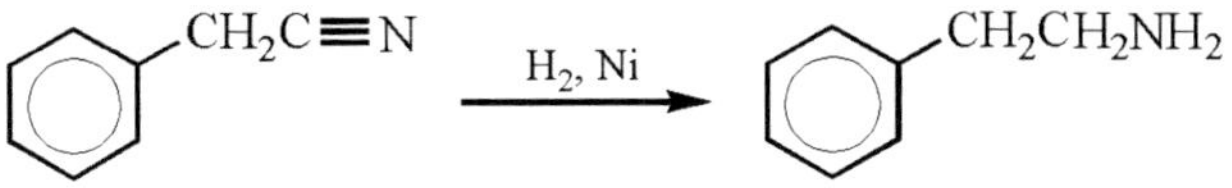

Phenylacetonitrile 2-Phenylethylamine

4. By reductive amination

When aldehydes and ketones are reduced in the presence of ammonia or amines, amines are formed.

$$CH_3CH_2C(CH_3){=}O + NH_3 \xrightarrow{H_2,\ Ni} CH_3CH_2CH(CH_3){-}NH_2$$

2-Butanone 1-Methylpropylamine

$$\underset{\text{Ethanal}}{CH_3CHO} + \underset{\text{Methylamine}}{CH_3NH_2} \longrightarrow \underset{\text{Ethylmethylamine}}{CH_3CH_2NHCH_3}$$

- *Key point* – In the above reaction, aldehydes and ketones react with ammonia to give a primary amine. If a primary amine is used instead of ammonia, the product is a secondary amine and a secondary amine gives rise to a tertiary amine.

5. By Hofmann degradation

When an aliphatic or aromatic amide is heated with hypobromite, an amine is formed.

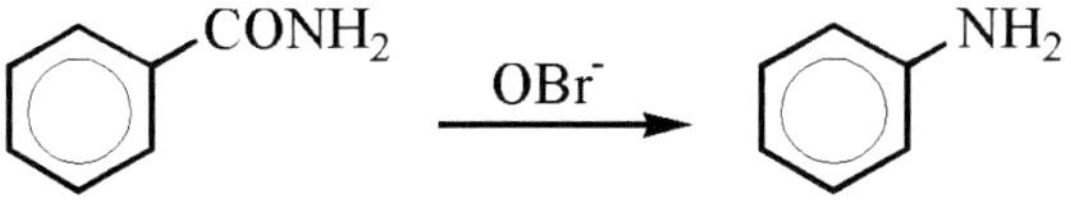

Benzamide Aniline

10.4 Reactions of amines

1. Formation of salts

Amines arc bases which thus combine with acids to form salts. These salts are alkyl substituted ammonium salts and named accordingly.

$$CH_3NH_2 + HCl \longrightarrow CH_3NH_3^{+}\,Cl^{-}$$

Methylamine — Methylammonium chloride

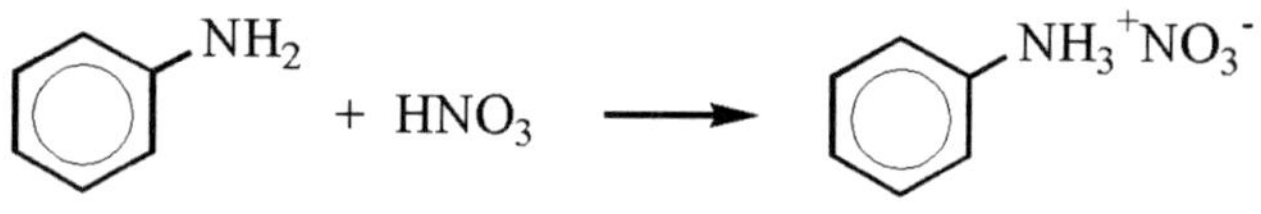

Aniline (Phenylamine) — Anilinium nitrate (Phenylammonium nitrate)

2. Reaction with alkyl halides

When amines are treated with alkyl halides, alkylation of amines takes place. A primary amine reacts with an alkyl halide to give a secondary amine, a secondary amine gives rise to a tertiary amine, and a tertiary amine to a quaternary ammonium salt (a quaternary ammonium salt is a tetra-alkyl substituted ammonium salt).

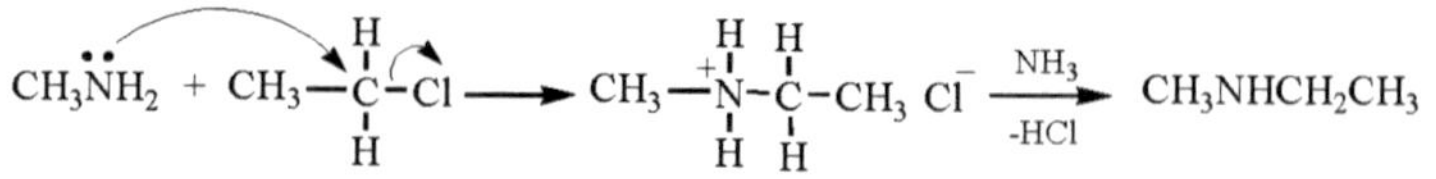

Methylamine (Primary amine) — Chloroethane — Intermediate salt — Ethylmethylamine (Secondary amine)

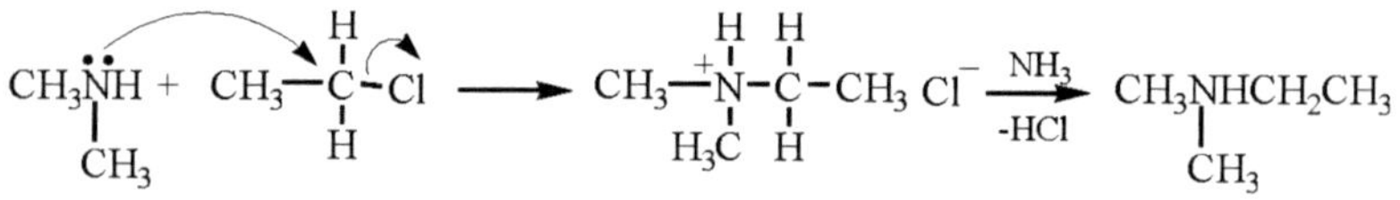

Dimethylamine (Secondary amine) — Chloroethane — Ethyldimethylamine (Tertiary amine)

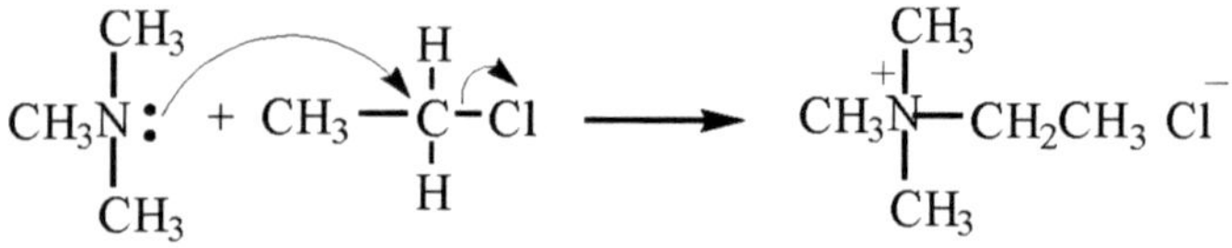

Trimethylamine (Tertiary amine) — Chloroethane — Ethyltrimethylammonium chloride (Quaternary ammonium salt)

► *Key point* – In the first two examples above, a proton is removed from the intermediate positive ammonium ion, but the quaternary ammonium salt has no hydrogen attached to the nitrogen and so the elimination of a hydrogen halide molecule does not take place.

3. Formation of amides

Primary and secondary amines react with acyl halides and acid anhydrides to form amides. Tertiary amines do not undergo similar reaction.

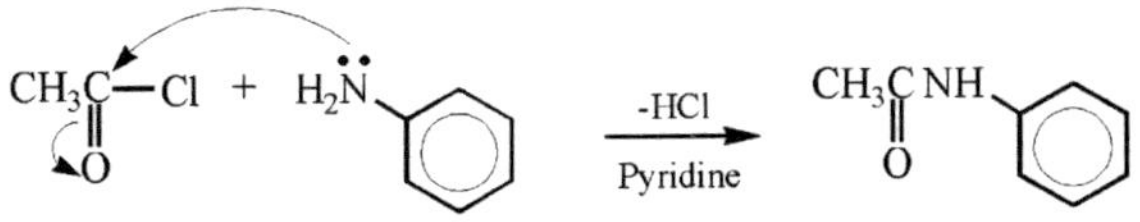

Ethanoyl chloride Aniline N-Phenylethanamide (N-Phenylacetamide)

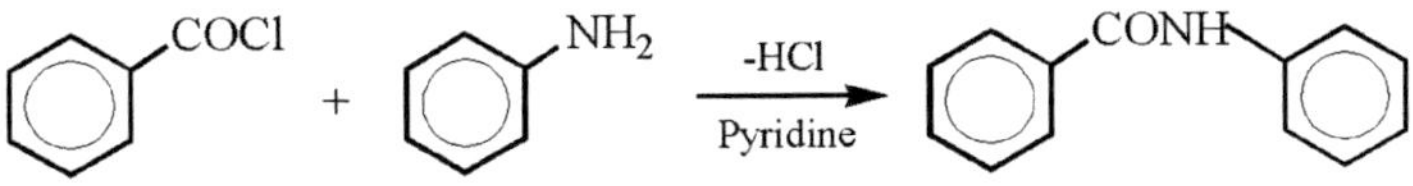

Benzoyl chloride Aniline N-Phenylbenzamide

▶ *Key point* – The reaction starts with the nucleophilic attack of the amine on the acyl carbon to form an intermediate, followed by the elimination of a molecule of hydrogen halide from the intermediate. Since a tertiary amine has no hydrogen attached to the nitrogen atom, a similar reaction does not take place.

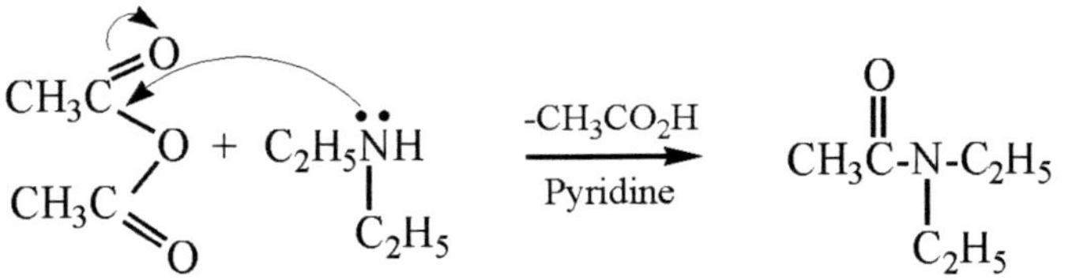

Ethanoic anhydride (Acetic anhydride) Diethylamine N,N-Diethylethanamide (N,N-Diethylacetamide)

4. Reaction with nitrous acid

Aliphatic primary amines react with nitrous acid (prepared by mixing sodium nitrite solution and hydrochloric acid) to form unstable intermediates which decompose to give alcohols and alkenes, together with nitrogen gas and water:

$$C_2H_5NH_2 + HONO \longrightarrow C_2H_5OH + N_2 + H_2O$$

Ethylamine Nitrous acid Ethanol

Aromatic primary amines form diazonium salts with nitrous acid at temperatures below 5°C. Diazonium salts are useful intermediates in the preparation of many classes of compounds.

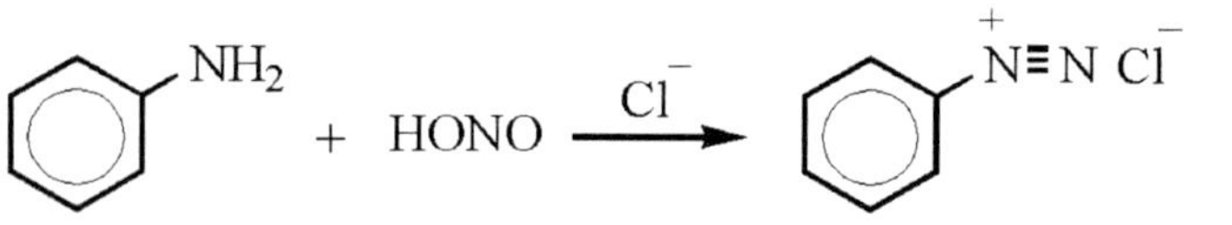

Aniline — Benzene diazonium chloride

Secondary amines react with nitrous acid to form N-nitrosoamines, which are yellow oily compounds.

$$(CH_3)_2NH + HONO \longrightarrow (CH_3)_2N{-}N{=}O + H_2O$$

Dimethylamine — N-Nitrosodimethylamine

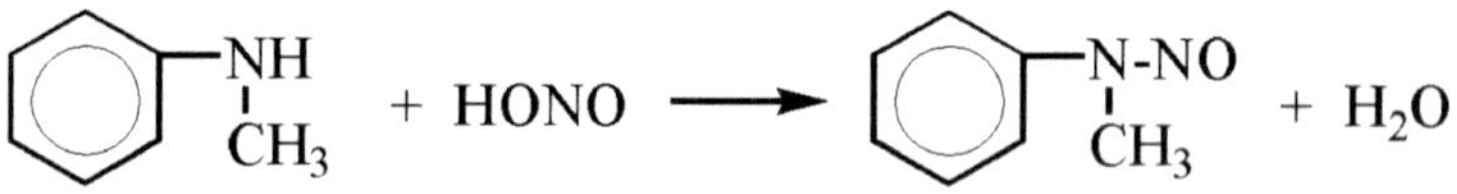

N-methylaniline — N-methyl-N-nitrosoaniline

Tertiary amines react with nitrous acid to give salts:

$$(CH_3)_3N + HNO_2 \longrightarrow (CH_3)_3NH^{+}NO_2^{-}$$

Trimethylamine — Trimethylammonium nitrite

5. Hofmann elimination

A quaternary ammonium hydroxide decomposes on heating to give an alkene, a tertiary amine and water.

$$CH_3CH_2CH_2CH_2N(CH_3)_3^{+}\,OH^{-} \longrightarrow CH_3CH_2CH{=}CH_2 + (CH_3)_3N + H_2O$$

1-Butyltrimethylammonium hydroxide — 1-Butene — Trimethylamine

10.5 Quaternary ammonium compounds

A quaternary ammonium ion contains a nitrogen bonded to four alkyl or aryl groups. A quaternary ammonium salt consists of a positively charged quaternary ammonium ion and an anion, and behaves like a salt. They are soluble in water and decompose on heating. Note the formulae and names of the quaternary ammonium compounds given below.

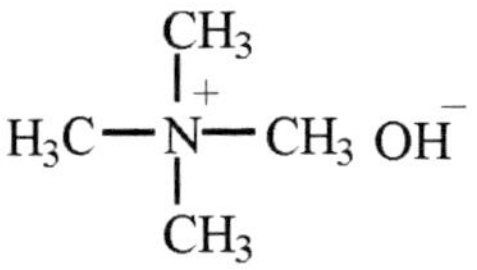

Tetramethylammonium hydroxide

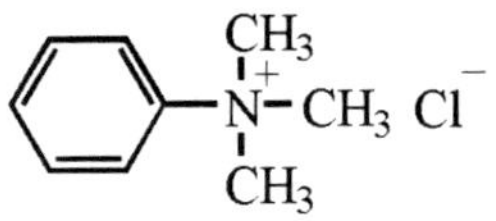

Trimethylphenylammonium chloride

10.6 Diazonium salts

Aryl diazonium salts are prepared by mixing cold solutions of primary arylamines and nitrous acid at temperatures below 5°C. At higher temperatures, they tend to decompose fast.

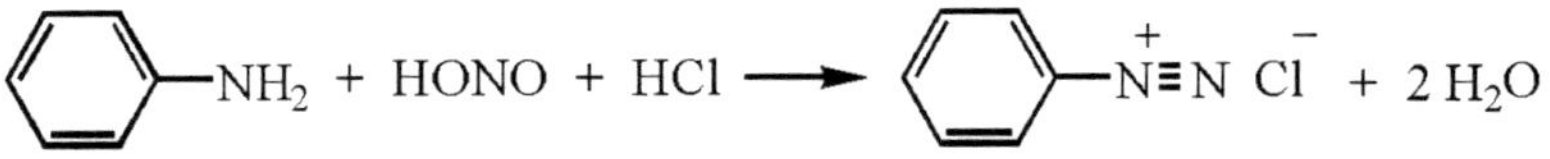

Aniline   Nitrous acid   Benzene diazonium chloride

Aryl diazonium salts can be converted into a number of compounds containing other functional groups, by substituting the diazonium group. The diazonium group can be replaced by -H, -OH, -F, -Cl, -Br, -I, -CN, $-NO_2$ and -Ar. Diazonium salts are thus useful intermediates of synthetic importance.

Reactions of diazonium salts

1. Replacement of the diazonium group by -Cl, -Br and -CN

Sandmeyers reaction allows the conversion of a diazonium salt into its chloro, bromo or cyano derivative by treating it with cuprous chloride, bromide or cyanide respectively.

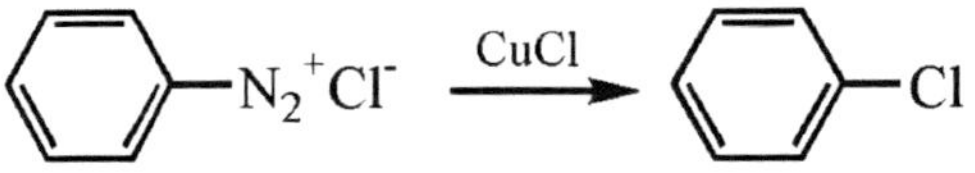

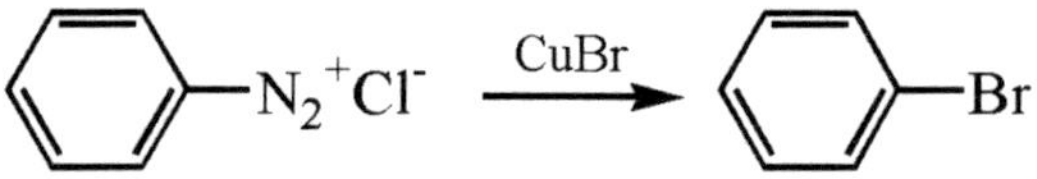

2. Replacement of the diazonium group by -I

Heating with potassium iodide gives rise to the iodine substituted derivative.

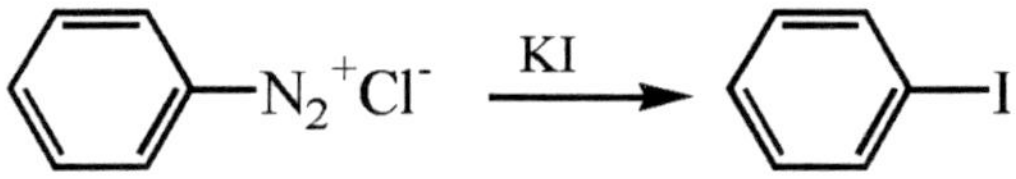

3. Replacement of the diazonium group by -OH

An acidified solution of a diazonium salt produces a corresponding phenol when heated.

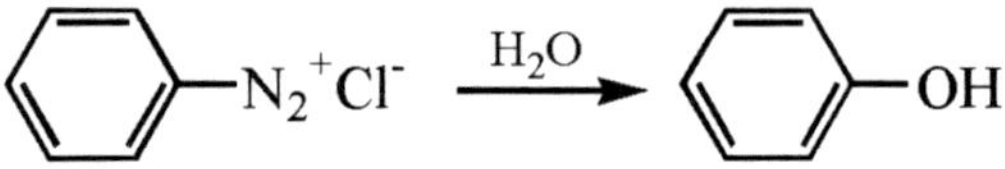

4. Replacement of the diazonium group by -H

On heating with hypophosphorous acid (H_3PO_2), the diazonium group is substituted by a hydrogen atom. This overall reaction is called **deamination**. Phenyl diazonium chloride on heating with H_3PO_2 gives benzene.

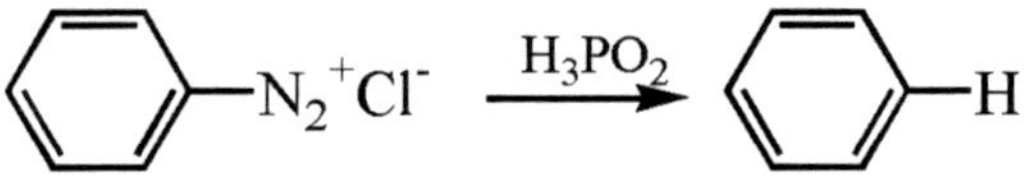

5. Diazotisation and the dye formation reaction

An aromatic primary amine may be diazotised and coupled with reactive phenols and aromatic tertiary amines to form coloured products, **dyes**.

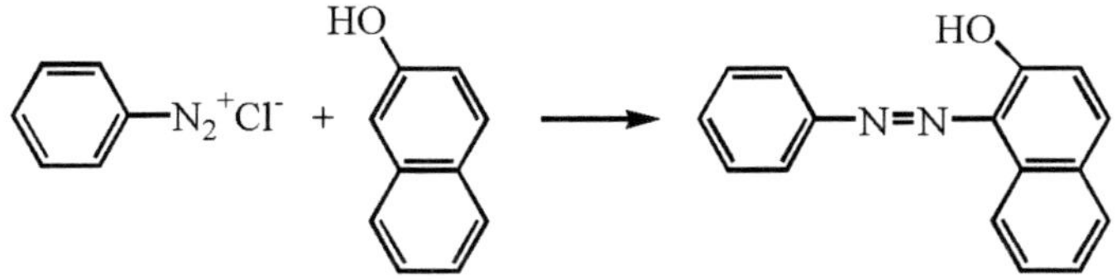

2-Naphthol A dye

Tutorial: helping you learn

Progress questions

1. Draw the structural formulae of the following compounds:

   (a) Triethylamine
   (b) Diphenylamine
   (c) Phenylammonium chloride
   (d) 4-Nitroaniline.

2. Give the IUPAC names of the following compounds:

   (a) $CH_3CH_2CH_2NHCH_2CH_2CH_3$

   (b) 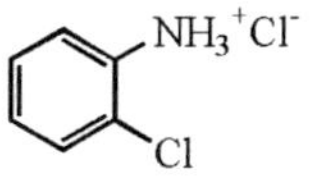

   (c) 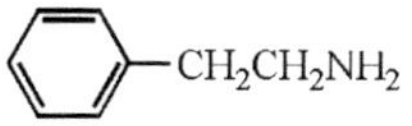

   (d) 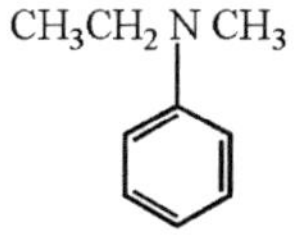

3. Which is more basic?

   (a) $CH_3CH_2NH_2$ or CH_3CONH_2

   (b) 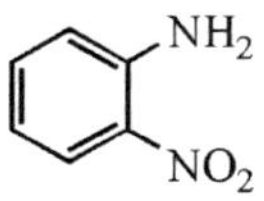 or 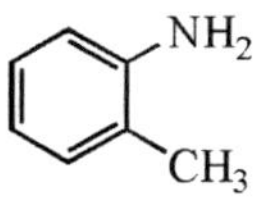

   (c) 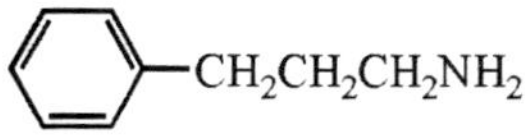 or 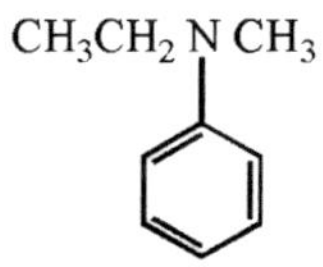

4. Suggest reagents and conditions for the following conversions (some of the conversions require more than one step):

   (a) Benzoic acid to benzamide.
   (b) Nitrobenzene to chlorobenzene.
   (c) Ethylamine to diethylamine.
   (d) Aniline to phenol.
   (e) Aniline to N-phenylacetamide.
   (f) Chlorobenzene to 2-chloroaniline.

5. Suggest reagents and conditions for the following conversions (a) - (d):

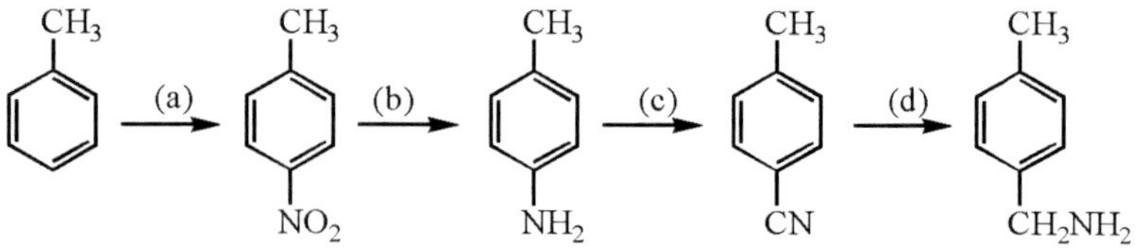

Discussion points

Penicillin antibiotics have a structural unit common to them, part of which is a lactam ring. **Lactams** are cyclic amides. Investigate the chemistry of the formation of lactams.

Practical assignment

Morphine and codeine, two powerful analgesics, belong to a family of compounds called **alkaloids**. These nitrogen-containing, basic compounds are distributed widely in the plant kingdom and possess pharmacological properties. Find the names of some common alkaloids and their plant origin.

11

Organic Synthesis

One-minute summary – Many compounds are synthesised by organic chemists for everyday use from simple and easily available materials. The strategy of converting simple compounds into more complex ones is the main theme of this chapter. First of all, the target molecule is disconnected into imaginary fragments. Next, compounds which can produce these fragments, directly or otherwise, are thought of, and a synthetic scheme proposed. Some of the points to be considered in selecting a synthetic route are: the availability of starting materials, yield of product and separation and purification of the required product from the reaction mixture. This chapter will cover:

- the terms target molecule, precursor and synthon
- retrosynthetic analysis
- synthesis: C-C bond formation
- functional group interconversion
- the use of protecting groups in organic synthesis.

11.1 Synthesis of organic molecules

Of the hundreds of thousands of organic compounds known to us, only a small fraction comes from nature, most being artificially synthesised. Organic chemists synthesise pharmaceuticals, agrochemicals, plastics, synthetic fibres and numerous other substances for everyday use. They employ different strategies in planning synthesis.

The required compound is called the '**target molecule**'. The target molecule is synthesised starting from simple and commercially available substances through one or a number of steps. A **precursor** is a substance that is converted into a target molecule.

11.2 What is retrosynthesis?

Retrosynthesis or retrosynthetic analysis is the reverse of a chemical reaction. It is the process of reasoning backwards from the target molecule to the starting material(s). In a retrosynthetic analysis, how a target molecule can be **disconnected** into precursors is considered. The relation between a target molecule and precursors is shown using an open arrow, as below,

$$\text{Target molecule} \Longrightarrow \text{Precursors}$$

In the process of synthesis, precursor(s) are converted into a target molecule.

$$\text{Precursor/s} \longrightarrow \text{Target molecule}$$

How do we plan a synthesis? Let us consider the preparation of propyne from ethyne as a simple example. First consider the retrosynthetic analysis.

Retrosynthetic analysis

$$\underset{\text{Propyne}}{H-C\equiv C-CH_3} \Longrightarrow \underset{\text{Sodium ethynide}}{H-C\equiv C^{-}\ Na^{+}} + \underset{\text{Bromomethane}}{CH_3-Br}$$

Then, give the reaction scheme for the synthesis using the precursors.

Synthesis

$$\underset{\text{Ethyne}}{H-C\equiv C-H} \xrightarrow{NaNH_2,\ liq.NH_3} \underset{\text{Sodium ethynide}}{H-C\equiv \bar{C}\ Na^{+}} \xrightarrow{CH_3Br} \underset{\text{Propyne}}{H-C\equiv C-CH_3}$$

In the above example, a C-C bond is formed. This can be shown in a different way in the retrosynthetic analysis:

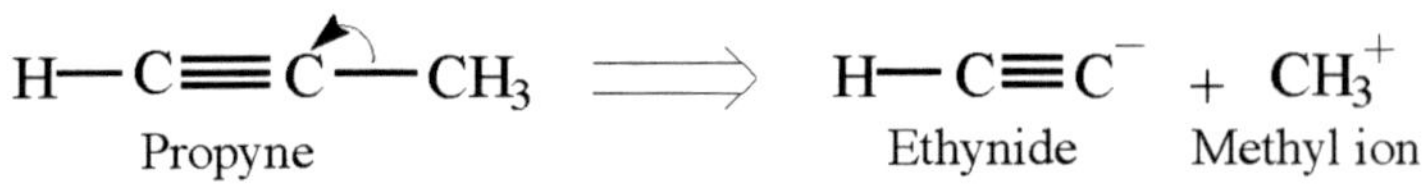

The fragments ethynide (acetylide) anion and methyl cation are called **synthons**. Synthons are usually cations or anions. Since synthons are normally not available directly, their synthetic equivalents are used as precursors. In this case, ethyne with sodamide in liquid ammonia provides ethynide ion and bromomethane is the synthetic equivalent of a methyl ion.

Let us take another example, the preparation of cyclohexylmethanol using a Grignard reagent. As a primary alcohol, it is formed by the reaction of methanal (formaldehyde) with a suitable Grignard reagent. Here again, a C-C bond is constructed.

Retrosynthetic analysis

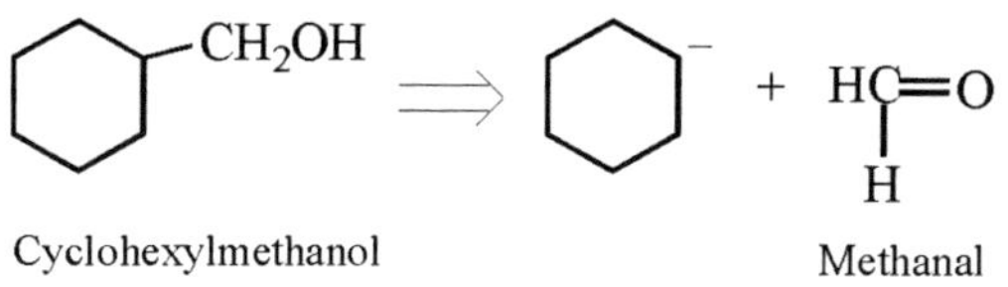

The precursors for the synthesis are methanal and cyclohexyl magnesium bromide (the latter to provide the anion).

Synthesis

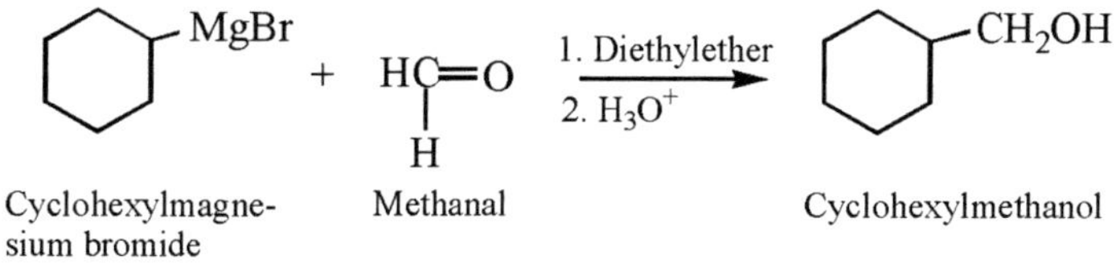

11.3 Construction of a C-C bond

In the examples given, it can be seen that the construction of a C-C bond is of considerable importance during synthesis. We have come across a number of reactions involving C-C bond formation. Some of these reactions are summarised here.

Summary of reactions

1. The reaction of ethynide (acetylide) ion with an alkyl halide (Section 6.5). Example:

$$H{-}C{\equiv}C^{-}\,Na^{+} + CH_3CH_2Br \longrightarrow H{-}C{\equiv}C{-}CH_2CH_3$$

Sodium ethynide Ethyl bromide 1-Butyne

2. The addition of hydrogen cyanide to aldehydes and ketones (Section 8.4). Example:

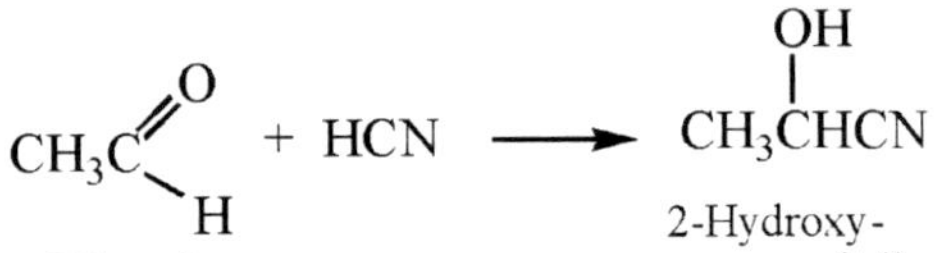

Ethanal 2-Hydroxy-propanenitrile

3. The reaction of an alkyl halide with a cyanide (Section 9.4). Example:

$$CH_3CH_2CH_2Br \xrightarrow{CN^-} CH_3CH_2CH_2CN$$

1-Bromopropane Butanenitrile

4. The addition of CO_2 on Grignard reagents (Section 9.4). Example:

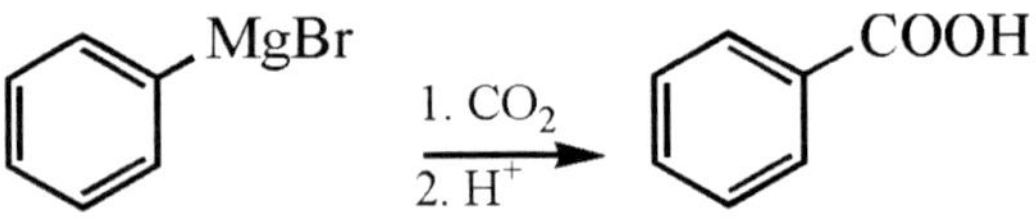

Phenyl magnesium bromide Benzoic acid

5. The reaction of Grignard reagents with aldehydes and ketones (Section 8.5). Example:

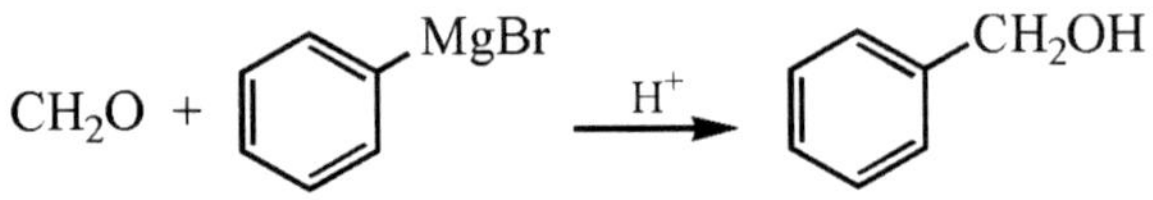

Methanal Phenyl magnesium bromide Benzyl alcohol

6. The Aldol reaction (Section 8.7). Example:

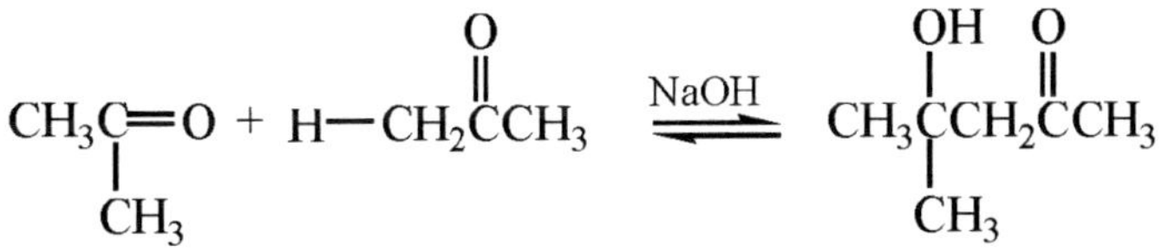

Propanone Propanone 4-Hydroxy-4-methyl pentan-2-one

7. The Wittig reaction (Section 8.5). Example:

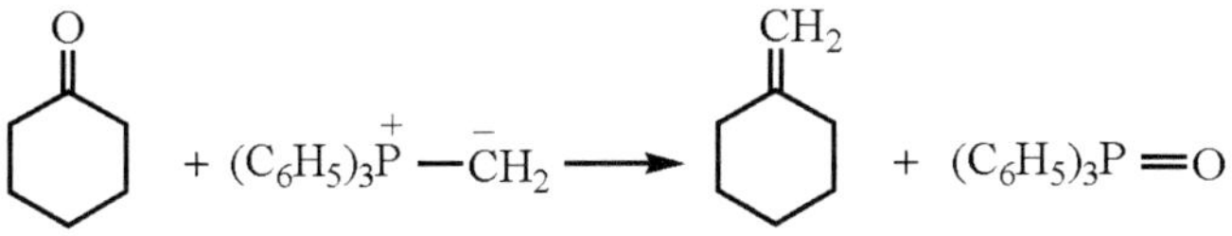

Cyclohexanone A phosphorane Methylenecyclohexane

11.4 Functional group interconversion (FGI)

An important method of organic synthesis is functional group interconversion. In earlier chapters we have studied many examples of the conversion of one functional group to another. More examples are discussed in the following sections.

11.5 Alkene synthesis

Alkenes can be prepared by:

1. The dehydration of alcohols (Section 7.5)

$$\underset{\text{2-Proanol}}{CH_3CHOHCH_3} \xrightarrow{-H_2O} \underset{\text{Propene}}{CH_3CH{=}CH_2}$$

2. The dehydrohalogenation of halogenoalkanes (Section 6.2)

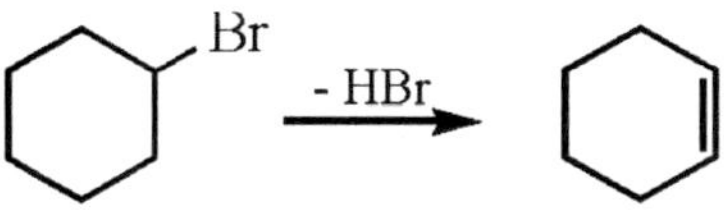

Cyclohexyl bromide Cyclohexene

3. The dehalogenation of vicinal dihalides (Section 6.2)

$$\underset{\text{2,3-Dichlorobutane}}{CH_3CHClCHClCH_3} \xrightarrow{Zn} \underset{\text{2-Butene}}{CH_3CH{=}CHCH_3}$$

4. The hydrogenation of alkynes (Section 6.2)

$$\underset{\text{2-Butyne}}{CH_3C{\equiv}CCH_3} \xrightarrow{\text{Catalyst}} \underset{\text{2-Butene}}{CH_3CH{=}CHCH_3}$$

5. The Wittig reaction (Section 8.5 and 11.3)

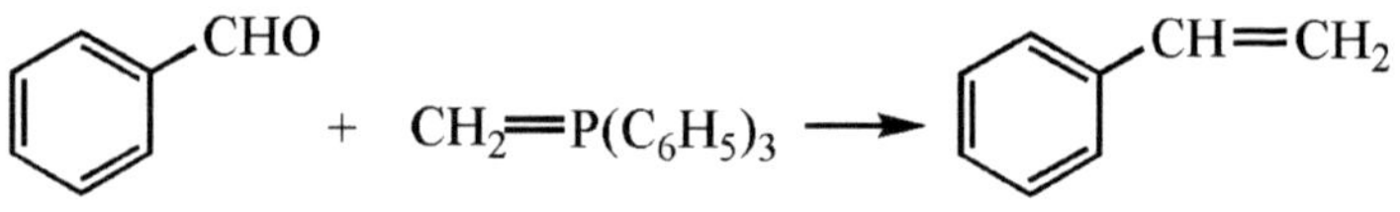

Benzaldehyde Phenylethene

11.6 Oxidation reactions of alkenes and alcohols

Alkenes are oxidised by cold, dilute, aqueous potassium permanganate to diols (Section 6.3). For example, potassium permanganate oxidises cyclohexene to cyclohexane-1,2-diol.

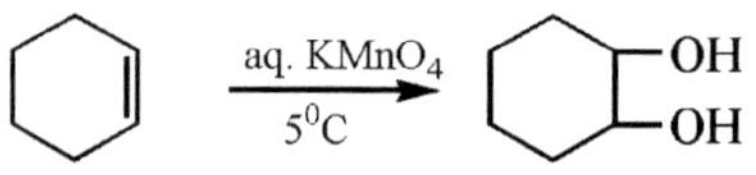

Cyclohexene Cyclohexane -1,2-diol

Ozone, O_3, can oxidise an alkene, cleaving the molecule at its double bond to produce two molecules with carbonyl groups (Section 8.3):

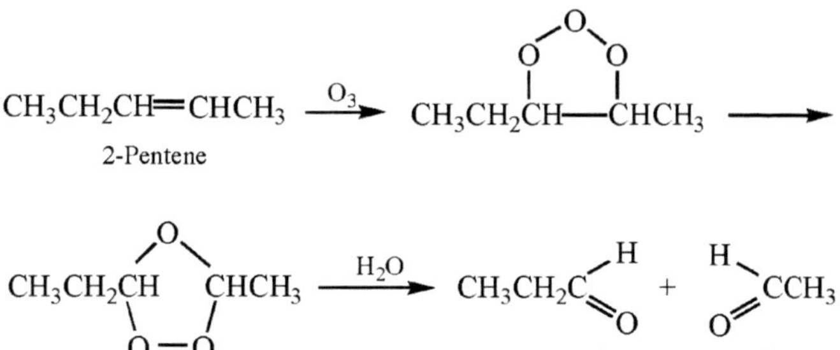

2-Pentene Propanal Ethanal

Alcohols undergo oxidation reactions with a variety of oxidising agents (Section 7.6). Potassium permanganate in an acid medium is a powerful oxidising agent and oxidises a primary alcohol into an aldehyde which is further oxidised to a carboxylic acid:

$$\underset{\text{1-Butanol}}{CH_3CH_2CH_2CH_2OH} \xrightarrow{KMnO_4,\ H^+} \underset{\text{Butanal}}{[CH_3CH_2CH_2CHO]} \longrightarrow \underset{\text{Butanoic acid}}{CH_3CH_2CH_2COOH}$$

Oxidising agents like pyridinium chlorochromate (PCC), oxidise primary alcohols to aldehydes in dichloromethane solvent:

$$\underset{\text{1-Hexanol}}{CH_3(CH_2)_4CH_2OH} \xrightarrow{PCC,\ CH_2Cl_2} \underset{\text{Hexanal}}{CH_3(CH_2)_4CHO}$$

Secondary alcohols are oxidised to ketones:

$$\underset{\text{2-Butanol}}{CH_3CH_2CHOHCH_3} \xrightarrow{[O]} \underset{\text{Butanone}}{CH_3CH_2COCH_3}$$

11.7 Reduction reactions of carbonyl compounds

Aldehydes and ketones can be reduced to alcohols by a number of methods (Section 8.4). An important synthetic route involves the use of sodium borohydride or lithium aluminium hydride. An aldehyde is reduced to a primary alcohol and a ketone to a secondary alcohol, usually with a high product yield. Study the examples given below.

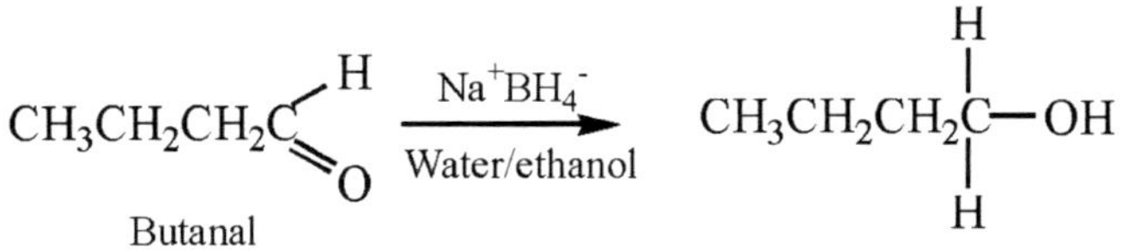

Butanal
(an aldehyde)

1-Butanol
(a primary alcohol)

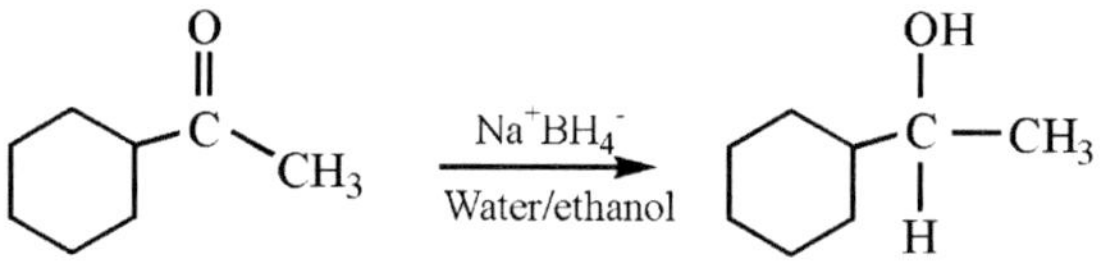

Cyclohexylmethylketone (a ketone)

1-Cyclohexylethanol (a secondary alcohol)

11.8 Protecting groups and strategy in organic synthesis

During the course of organic synthesis, we sometimes need to protect a functional group so that it does not interfere with the formation of another group or another conversion. This strategy of protecting functional groups is very important in organic synthesis.

Consider the preparation of 3-oxocyclohexylmethanol from 3-bromocyclohexanone.

3-Bromocyclo-hexanone

3-Oxocyclohexyl-methanol

The synthesis can be carried out by converting the bromocompound to the corresponding Grignard reagent, and then treating it with methanal. Since the carbonyl group reacts with the Grignard reagent, the synthesis cannot be done without protecting the carbonyl function. It is done by converting the ketone into a cyclic ketal which will not interfere with the Grignard reagent preparation. At the end, the protective group can be removed by hydrolysis with aqueous acid.

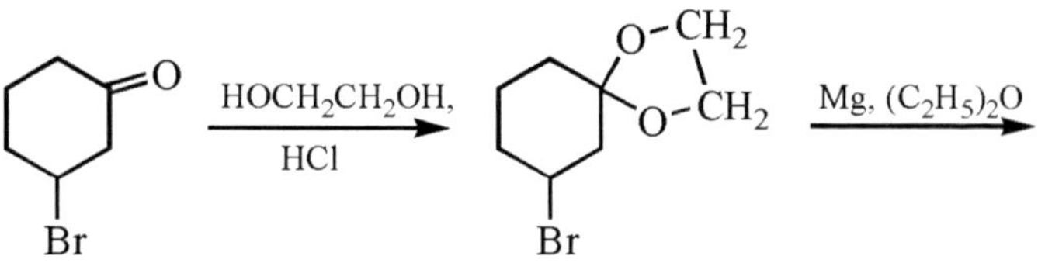

3-Bromocyclo-hexanone

Cyclic ketal

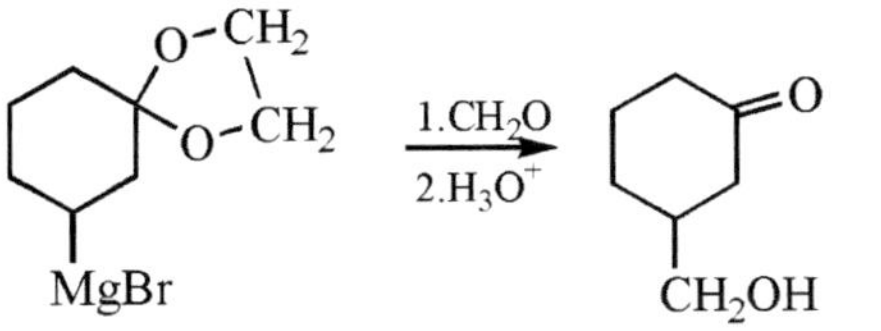

3-Oxocyclohexyl-methanol

Tutorial: helping you learn

Practice question

Outline a retrosynthetic analysis scheme and a plan for the synthesis of the following target molecules:

(a) 4-Methylacetophenone
(b) Benzylamine.

Progress questions

1. Describe a retrosynthetic analysis scheme for each of the following:

   (a) 2-Pentyne
   (b) 2-Phenylethanol
   (c) N-phenylbenzamide.

2. How would you plan the synthesis of each of the following?

   (a) 1-Hexyne
   (b) 2-Hexyne
   (c) 3-Methyl-1-pentene
   (d) 2-Pentanol
   (e) *cis*-1,2-cyclopentanediol
   (f) *trans*-2-pentene.

12

Analysis and Spectroscopy

One-minute summary – Spectroscopic analysis plays an important role in the determination of structure of organic molecules. In this chapter, we will briefly discuss infrared (IR) spectroscopy, ultraviolet (UV) spectroscopy, mass spectrometry and proton nuclear magnetic resonance (^{1}H NMR) spectroscopy. IR, UV and NMR spectroscopic methods involve the use of electromagnetic radiation of different ranges of wavelengths. The IR spectrum of a compound gives information about its functional groups, the UV spectrum about conjugate double bond system, the NMR spectrum about the carbon and hydrogen framework of the molecule, and the mass spectrum about its molecular weight. In this chapter, we will discuss:

- electromagnetic radiation
- wavelength, wavenumber and frequency
- the use of IR, UV and NMR spectroscopy as tools in organic analysis
- mass spectrometry and molecular weight determination.

12.1 Introduction to electromagnetic radiation

Visible light forms part of the **electromagnetic spectrum**. So do γ-rays, X-rays, infrared rays, ultraviolet rays and radio waves. Consider this entire spectrum as waves, travelling at a constant speed (the speed of light). These rays are called **electromagnetic radiation** since they have wave properties which involve electric and magnetic forces. The electromagnetic spectrum is arbitrarily divided into regions as shown in Figure 12.1.

Wavelength (λ) is the length of one complete wave, from trough to trough or from crest to crest and it is expressed in centimetres or metres. The visible region forms only a small part of the electromagnetic spectrum, of wavelength from about 4×10^{-7} m to 8×10^{-7} m (or 400 nm to 800 nm). Note that 1 nm = 10^{-9} m.

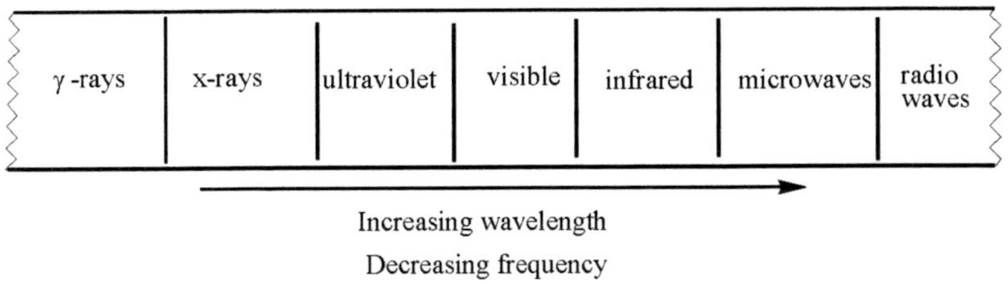

Figure 12.1. The electromagnetic spectrum.

Different rays have different wavelengths. But all rays travel at the same velocity in a vacuum. This is called the **velocity of light** (c), and its value is 2.998×10^{10} cm s^{-1}.

The reciprocal of wavelength is the **wavenumber** ($\bar{\nu}$), which is the number of waves per centimetre.

$$\bar{\nu} = \frac{1}{\lambda} \text{ cm}^{-1}$$

Frequency (ν) is the number of wave cycles that passes through a fixed point in unit time (usually per second). The unit is cycles per second or cps or hertz (Hz).

Wavelength and frequency are related by the equation

$$\lambda = \frac{c}{\nu}$$

where λ = wavelength in cm
c = velocity of light ($\sim 3 \times 10^{10}$ cm s^{-1})
ν = frequency in Hz

Electromagnetic radiation can also be considered as a stream of particles called **photons**. The energy of a photon is quantized and is related to the frequency of radiation by the equation $E = h\nu$, where

E = energy in kJ mol^{-1}
h (*Planck's constant*) = 3.99×10^{-13} kJ s mol^{-1}

Since

$$\nu = \frac{c}{\lambda}$$

$$E = h\frac{c}{\lambda}$$

It follows from the above equation that the energy of a radiation and its wavelength are related. The shorter the wavelength, the higher the energy. Or, the energy of a radiation is directly proportional to its frequency. Higher frequency radiation has higher energy.

12.2 Infrared spectroscopy

The infrared region of the electromagnetic spectrum covers 2×10^{-7} m to 7.8×10^{-3} m wavelength. **Infrared spectroscopy** is a technique whereby the amount of infrared radiation a substance absorbs or transmits is measured and plotted on a graph to give an **infrared spectrum**. Infrared spectroscopy is an important tool in the identification of functional groups in organic compounds.

Molecules are not rigid bodies; the bonds in molecules can stretch and bend, or vibrate like a spring. The energies for these vibrations are quantized. This means that a specific amount of energy is absorbed to effect the vibration of a particular bond, and to thus raise it to a higher energy level. The energy required for these vibrations in covalent molecules ranges from about 8 to 42 kJ mol^{-1}. This range of energy is available when a substance is irradiated with infrared light of wavelength ranging from 2.5×10^{-6} m to 2.5×10^{-5} m, which is equivalent to wavenumbers from 4000 cm^{-1} to 400 cm^{-1}. This corresponds to the middle portion of the infrared region.

Infrared spectrometers can be used easily to obtain the infrared spectrum of a compound. A liquid sample can be used directly, and is placed in between two sodium chloride discs to form a thin film. A solid sample is ground with a drop of liquid paraffin (Nujol) to form a paste (Nujol mull) or can be dissolved in a solvent like chloroform or carbon tetrachloride with minimum infrared absorption. Sometimes a small sample of the substance is ground with potassium bromide and compressed into a disc before being used. Sodium chloride and potassium bromide are transparent to infrared light while glass is not.

The sample is placed in the spectrometer and irradiated with infrared light of wavenumber 4000 to 400 cm^{-1}. The sample absorbs infrared light of some frequencies while others are transmitted. The spectrometer automatically compares the intensity of the absorbed light with that of a reference beam, and plots the percentage of the transmitted light against the wavenumber on a chart. Each downward peak represents IR absorption of a specific wavenumber. A typical IR spectrum is given in Figure 12.2.

Functional group	*Wave number*	*Intensity*	*(cm^{-1})*
Alkanes			
C-H stretch	2850-3000	s (strong)	
Alkenes			
=C-H stretch	3020-3100	m (medium)	
C=C stretch	1650-1670	m	
C-H out of plane bend	910, 990	s, s	
Alkynes			
C-H stretch	3300	s	
C-H bend	600-700	s	
C≡C stretch	2100-2140	s	
Aromatic compounds			
C-H stretch	3000-3100	m	
Alkyl halides			
C-Cl	600-800	s	
C-Br	500-600	s	
C-I	500	s	
Alcohols			
O-H stretch	3200-3600	s	
C-O stretch	1050-1150	s	
Carbonyl compounds			
C=O stretch in ketones	1715	s	
in aldehydes	1725	s	
in carboxylic acids	1700-1725	s	
in esters	1730-1750	s	
in amides	1680-1700	s	
Carboxylic acids			
O-H stretch	2500-3100	s, broad	
Nitriles			
C≡N stretch	2210-2260	m	
Amines			
N-H stretch	3300-3500	m	
C-N stretch	1030-1230	m	

Table 12.1. Characteristic infrared absorption.

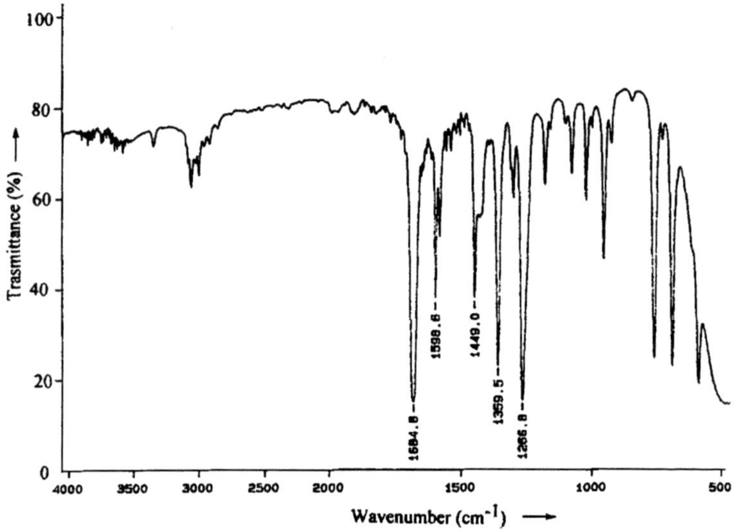

Figure 12.2. IR spectrum of acetophenone.

The molecules of a substance are **infrared active** only if the frequency of radiation matches the frequency of vibration of the bonds. If the bonds are polar, the intensity of absorption is greater i.e. the greater the polarity, the higher the absorption intensity. Thus, molecules like carbon monoxide and methane, or bonds like C-H, C=O, N-H and C-N are infrared active while molecules like H_2 and Cl_2 are infrared inactive.

The infrared spectrum of a compound can be complex and difficult to interpret completely. This is because there can be very many vibrations of the bonds which gives rise to numerous absorption peaks. But most functional groups give rise to characteristic absorption peaks between 3500 and 1600 cm^{-1}. The region between 1600 and 400 cm^{-1} is called the **fingerprint region** since the infrared spectrum of a compound is unique in this region, and is different for different substances.

As mentioned above, most functional groups have absorption peaks between 3500 and 1600 cm^{-1}. For example, the C=O group shows absorption between 1690-1780 cm^{-1}, C=C at 1650-1670 cm^{-1} and O-H of alcohols at 3200-3600 cm^{-1}. There can be some variation in the position of the absorption peak of same functional group in different compounds. As an example, consider the carbonyl stretching absorption in ketones and esters. The carbonyl stretching is at 1685 cm^{-1} in acetophenone (Figure 12.2) and at 1747 cm^{-1} in ethyl acetate (Figure 12.3).

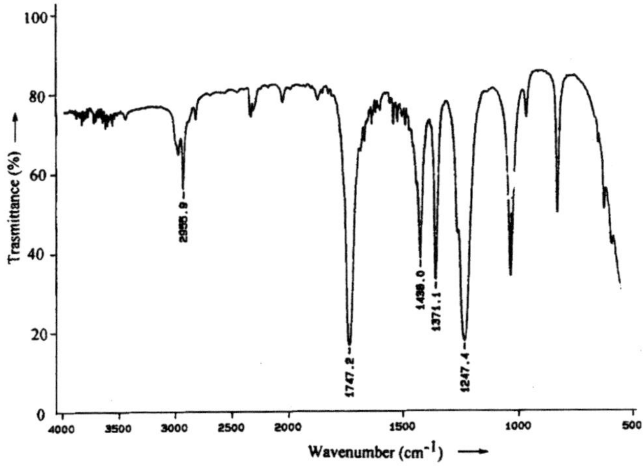

Figure 12.3. IR spectrum of ethyl acetate.

12.3 Ultraviolet spectroscopy

The ultraviolet (UV) region lies between visible light and X-rays on the electromagnetic spectrum, covering a region of wavelength from 10^{-8} m to 4×10^{-7} m (figure 12.1). Of this, the region from 2×10^{-7} m to 4×10^{-7} m (which is equal to 200-400 nm) is used by organic chemists for structural studies.

When organic compounds are subjected to UV radiation, some electrons absorb energy from the radiation, get excited and are promoted to higher energy levels. It can be seen that compounds containing conjugated double bonds and aromatic compounds absorb energy in the 200-400 nm region. So UV spectroscopy is mostly used to study conjugated multiple bond systems in molecules.

A sample of the compound under investigation is dissolved in a suitable solvent which does not absorb UV radiation in the region being analysed. Often, 95% aqueous ethanol is used as it is UV transparent down to 205 nm. Absolute ethanol is not used as it contains some benzene which is UV active. Cyclohexane is transparent down to 190 nm.

The solution of the sample is placed in a quartz cell and the solvent in an identical cell. Glass cells cannot be used as glass absorbs UV light. UV radiation of continuously varying wavelength is split into two beams, one being passed through the sample cell, and the other through the solvent cell. The transmitted radiation from both cells are compared electronically and the result is plotted as a graph of wavelength (on the x-axis) versus the percentage radiation absorbed (on the y-axis).

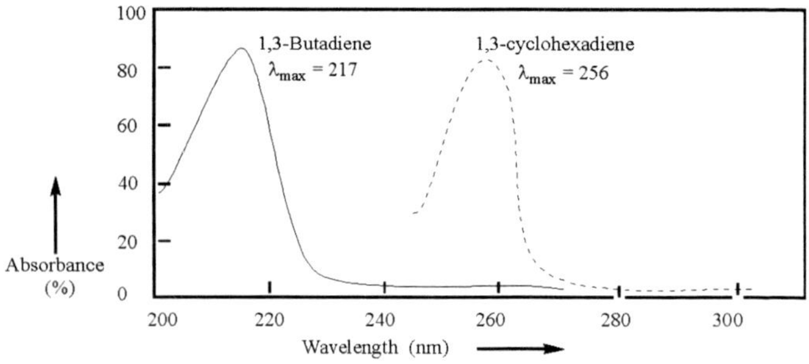

Figure 12.4. UV spectrum.

As mentioned earlier, when an organic compound containing conjugated double bonds is subjected to UV radiation, the π electrons absorb radiation of a certain wavelength and are promoted to higher energy levels. The energy absorbed appears on the graph as a broad peak. In the absorption spectrum, the wave length corresponding to the highest point of the peak is noted, and this is called λ_{max}. The value of λ_{max} of 1,3-butadiene is 217 nm and that of 1,3-cyclohexadiene is 256 nm.

The UV spectra of many organic compounds are recorded in the literature. The structural unit responsible for the UV absorption peak in a compound is called a **chromophore**. The absorption spectrum of an unknown compound is compared with spectra of compounds of known chromophores in order to identify the part-structure of the compound. Ethene has a λ_{max} of 171 nm. Compounds with non-conjugated double bonds have λ_{max} values below 200 nm. Note that the λ_{max} value for 1,4-pentadiene, which does not have a conjugated double bond system is 178 nm. Compounds with conjugated multiple bond systems have peaks between 200 and 400 nm.

As the number of double bonds in a conjugated system increases, the value of λ_{max} increases. Compare the values for 1,3-butadiene and 1,3,5-hexatriene (Table 12.2). Benzene with its conjugated double bond system has a λ_{max} value of 254 nm. The conjugated multiple bond system may include C=O, C=N and C=S. The non-bonding electron pairs (lone pairs) of O, N and S may also contribute to extend the conjugation.

Substituents like a methyl group increase the λ_{max} value. The λ_{max}

values of 1,3-butadiene, 2-methyl-1,3-butadiene and 2,3-dimethyl-1,3-butadiene are 217, 220 and 227 nm respectively (Table 12.2).

Compound	Structure	λ_{max} (nm)
Ethene	$CH_2{=}CH_2$	171
1,3-Butadiene	$CH_2{=}CH\,CH{=}CH_2$	217
1,3,5-Hexatriene	$CH_2{=}CH\,CH{=}CH\,CH{=}CH_2$	258
2-Methyl-1,3-butadiene	$CH_2{=}C(CH_3)\,CH{=}CH_2$	220
2,3-Dimethyl-1,3-butadiene	$CH_2{=}C(CH_3){-}C(CH_3){=}CH_2$	227
1,3-Cyclohexadiene		256
Benzene		254
Acetone	$CH_3\,C(=O)\,CH_3$	280

Table 12.2. λ_{max} values.

12.4 Mass spectrometry

The molecular mass of a substance can be determined accurately using a mass spectrometer. Mass spectrometry differs from IR and UV spectroscopic methods since there is no absorption of electromagnetic radiation involved. When an organic substance is bombarded with high energy electrons, usually ~70 eV (electron-Volts), an electron may be knocked out of a molecule to produce a positively charged ion. This ion is called a **molecular ion** (M^+). The mass of a molecular ion is the same as that of the original molecule since the lost electron has a negligible mass. During the bombardment, some energy is transferred from the electron beam to the molecular ion. Thus, the molecular ion may then fragment into other positive ions or radicals as shown below:

$$A{-}B \text{ or } A{:}B + e \longrightarrow \underset{\text{molecular ion}}{A\bullet B^{+}} + 2e$$

$$A\bullet B^{+} \longrightarrow A\bullet + B^{+}$$

or

$$A\bullet B^{+} \longrightarrow A^{+} + B\bullet$$

In a mass spectrometer, positively charged ions are produced, separated by their **mass-to-charge ratio** (m/z ratio), detected and recorded on a chart to form a **mass spectrum**. The position of a peak in a mass spectrum represents the mass of a positive ion and the height of the peak, its relative abundance.

A sample of the organic substance under investigation is vaporised in the ionising chamber of the spectrometer at a very low pressure (10^{-6} to 10^{-7} torr). Bombarding it with a beam of high energy electrons produces positive ions. These positive ions are then accelerated by passing them through two electrically-charged metal plates or grids.

The accelerated beam is passed through a slit and then subjected to a magnetic field generated by an electromagnet, which bends the beam into a circular path. The radius of the path depends on the m/z ratio of the ions. Thus, ions with different m/z ratios are separated, and take different paths. Ions with the same m/z ratio are allowed to arrive at a detector through a slit, where they produce a current. The intensity of the current produced depends on the number of ions. The mass of the particles in the beam is obtained from the known strengths of the applied electric and magnetic fields. By changing the magnetic field strength, ions of other m/z ratios can successively be allowed into the detector and recorded.

Mass spectra are recorded in the form of a bar chart, in which relative intensity is plotted as the ordinate (y-axis) versus m/z as the abscissa (x-axis). The position of the peak denotes the mass. The most intense peak is called the **base peak** and is arbitrarily given an intensity of 100%. The heights of other peaks are proportional to their abundance in comparison with the base peak.

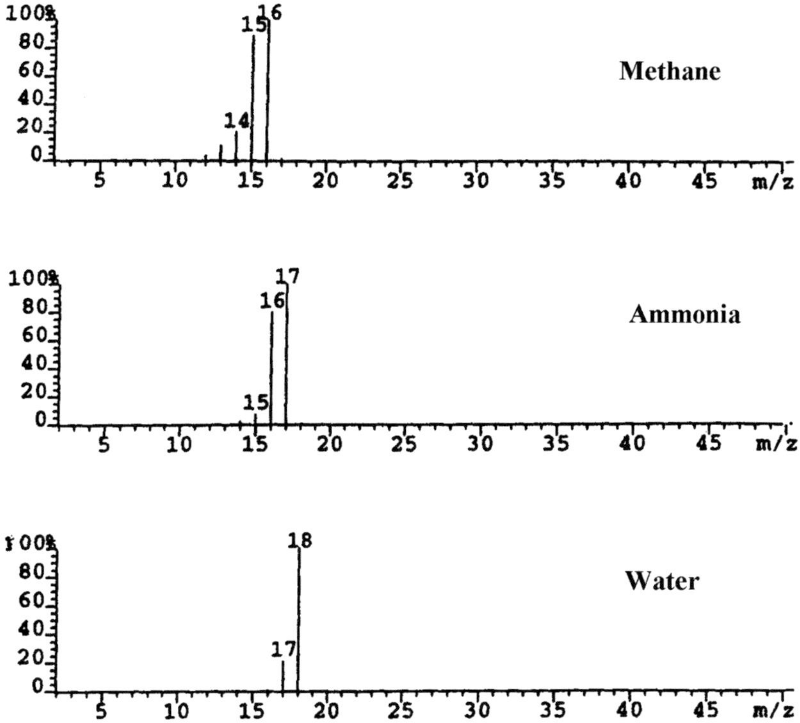

Figure 12.5. Mass spectra of CH_4, NH_3 and H_2O.

Molecular fragmentation

The mass spectrum of methane (figure 12.5) contains peaks at m/z 17 (M+1), 16 (M^+), 15,14,13,12 and 1. These are formed from the following ionisation or fragmentation steps.

$$CH_4 + e \longrightarrow \underset{m/z\ 16}{CH_4^{+\bullet}} + 2e$$

$$CH_4^{+\bullet} \longrightarrow \underset{m/z\ 15}{CH_3^{+}} + H\bullet$$

$$\text{And, } CH_4^{+\bullet} \longrightarrow CH_3\bullet + \underset{m/z\ 1}{H^{+}}$$

$$CH_3^{+} \longrightarrow \underset{m/z\ 14}{CH_2^{+\bullet}} + H^{\bullet}$$

$$CH_2^{+\bullet} \longrightarrow \underset{m/z\ 13}{CH^{+}} + H^{\bullet}$$

$$CH^{+} \longrightarrow \underset{m/z\ 12}{C^{+\bullet}} + H^{\bullet}$$

The M+1 peak often comes from an isotope of higher mass. For example, the peak at m/z 17 in the mass spectrum of methane is due to a ^{13}C or an ^{2}H in the molecular ion.

Different types of compounds have different patterns of fragmentation. Stability of the carbocation as well as stability of the free radical formed during fragmentation are important factors. In the mass spectrum of 2-methyl butane, the base peak is at m/z 43 which corresponds to a stable secondary cation, while the mass spectrum of its isomer, 2,2-dimethylpropane gives a base peak at m/z 57 corresponding to a tertiary cation.

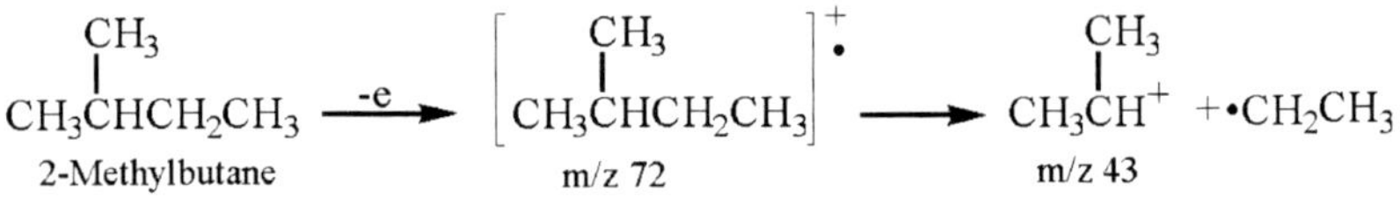

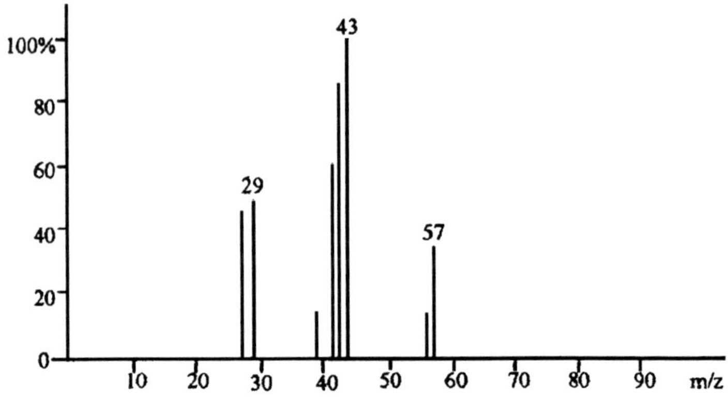

Figure 12.6. Mass spectrum of 2-methylbutane.

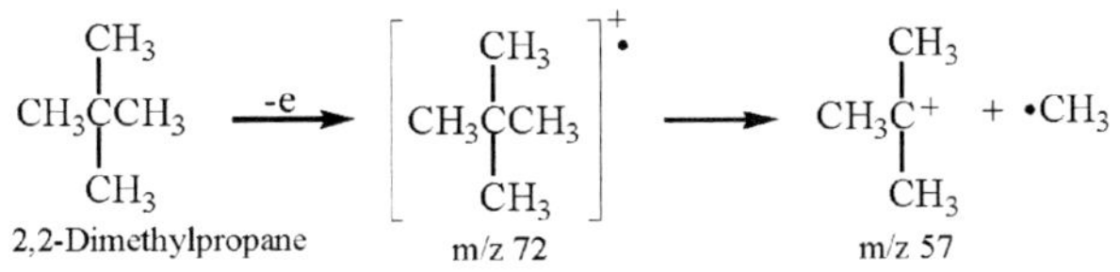

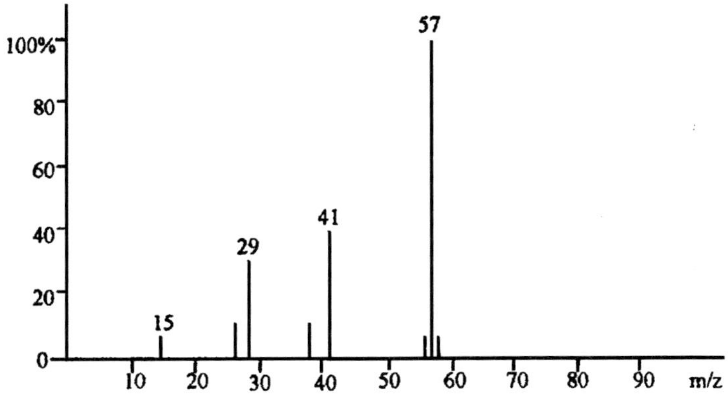

Figure 12.7. Mass spectrum of 2,2-dimethylpropane.

Some molecules are more prone to fragmentation than others. The base peak in the mass spectrum of benzene, C_6H_6, corresponds to its molecular ion, at m/z 78. Monosubstituted benzenes usually lose their substituents to give a cation ($C_6H_5^+$), of m/z 77. For many alkyl benzenes, the base peak is at m/z 91 which corresponds to the stable benzylic cation, $C_6H_5CH_2^+$ or tropylium ion, $C_7H_7^+$.

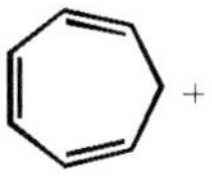

Tropylium ion

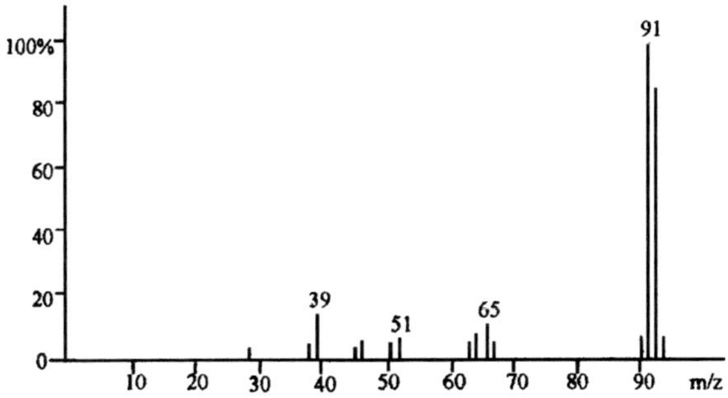

Figure 12.8. Mass spectrum of toluene.

12.5 Nuclear magnetic resonance spectroscopy

The nucleus of a ^1H atom spins about an axis, and since it is positively charged, it behaves like a small bar magnet. The spinning is random. When a compound containing ^1H is placed in between the poles of a powerful magnet, some nuclei are aligned with the applied magnetic field, and some are aligned against it. The energy level of a nucleus aligned with the magnetic field is lower than that of a nucleus aligned against it. If these nuclei are irradiated with electromagnetic radiation of appropriate frequency, the lower energy nuclei can absorb energy and flip into a higher energy state. This is called **resonance**. The energy absorbed is detected and recorded as a **signal** in an **NMR spectrum** using an NMR spectrometer.

Each ^1H nucleus in a molecule is surrounded by an electron cloud which creates a **local magnetic field** when placed in an external magnetic field. So the effective magnetic field experienced by a ^1H nucleus is less than the applied magnetic field because of the **shielding effect** of the local magnetic field. And since the shielding effect can vary for different ^1H nuclei in a molecule, their effective magnetic fields also vary. The effective magnetic field of nuclei is also altered by the presence of nearby protons. Thus, different ^1H nuclei in a molecule, with different electron environments, produce signals at different irradiation frequencies in an NMR spectrum.

A proton which is more shielded will absorb at a higher magnetic field strength than one which is less shielded. As an example, consider the NMR signals produced by the protons in chloroethane, CH_3CH_2Cl. The electron withdrawing inductive effect of chlorine reduces the electron density of the C-1 protons. These protons are thus less shielded than the C-2 protons. So C-2 protons (CH_3 protons) give signals more upfield compared to the signals produced by C-1 protons (CH_2Cl protons).

In an NMR spectrum, magnetic field strength is recorded along the base line. The applied field strength increases from left to right. Signals on the left of the spectrum are said to be at **downfield** and those on the right, **upfield**.

Chemical shift (δ)

Most chemical shifts have values between 0 and 10 ppm. The NMR spectrum is calibrated using tetramethylsilane (TMS), $(CH_3)_4Si$, as the reference compound, and the signal produced by the 12 equivalent protons of TMS is assigned an arbitrary δ value of 0.0 ppm. The

chemical shift is the position of a signal in the NMR spectrum and is expressed as parts per million (ppm) downfield from TMS.

The protons of most organic molecules we come across give rise to signals downfield from the TMS signal. In a molecule, equivalent protons have the same chemical shift whereas non-equivalent protons have different chemical shifts. For example, the six protons in propanone ($(CH_3)_2CO$) are equivalent and propanone gives a single signal at δ 2.1. In the NMR spectrum of methyl acetate, CH_3COOCH_3, there are two single peaks. The methyl group attached to the carbon gives a peak at δ 2.0 and the one attached to the oxygen at δ 3.6. The two methyl protons are non-equivalent.

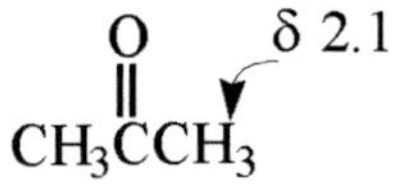

Acetone gives a six-proton singlet at δ 2.1

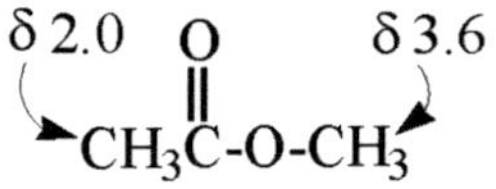

The two methyl groups of methyl acetate are non-equivalent

Type of proton	Structure	Chemical shift, δ ppm
Saturated primary	RCH_3	0.7 - 1.3
Saturated secondary	RCH_2R	1.2 - 1.4
Saturated tertiary	R_3CH	1.4 - 1.7
Alkyl iodide	RCH_2I	2.0 - 4.0
Alkyl bromide	RCH_2Br	2.5 - 4.0
Alkyl chloride	RCH_2Cl	3.0 - 4.0
Methyl ketone	$RCOCH_3$	2.1 - 2.4
Aryl methyl	$ArCH_3$	2.2 - 2.5
Alcohol	ROH	0.5 - 6.0
Amine	R_2NH	0.5 - 5.0
Aromatic	ArH	6.5 - 8.0
Aldehyde	RCHO	9.7 - 10.0
Carboxylic acid	RCO_2H	11.0 - 12.0

Table 12.3. Chemical shifts.

Spin-spin splitting

The nuclear spin of one atom can interact with the nuclear spin of a nearby atom. As a result, the absorption peak or signal splits to give multiple absorption peaks called **multiplets**. Protons that have *n* number of neighbouring protons show *n+1* peaks. For example, chloroethane, CH_3CH_2Cl produces two sets of peaks. The peaks at δ 3.59 corresponding to $-CH_2-$ are a quartet (i.e. four peaks) because its signal is split by three CH_3 protons. Similarly the peaks at δ 1.49 corresponding to the CH_3- protons are a triplet because of the coupling with the two CH_2 protons.

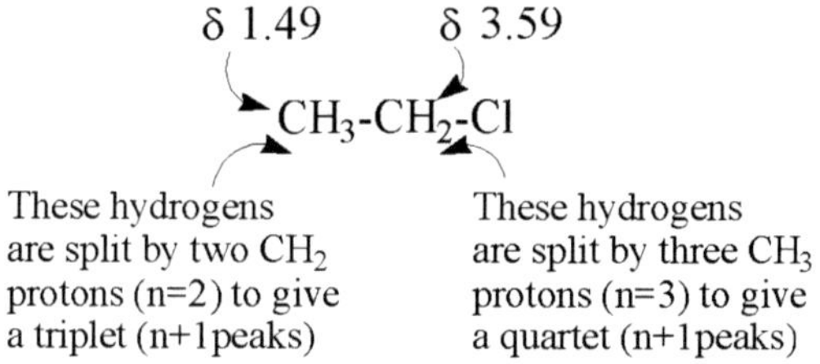

The spin-spin splitting is shown diagrammatically below.

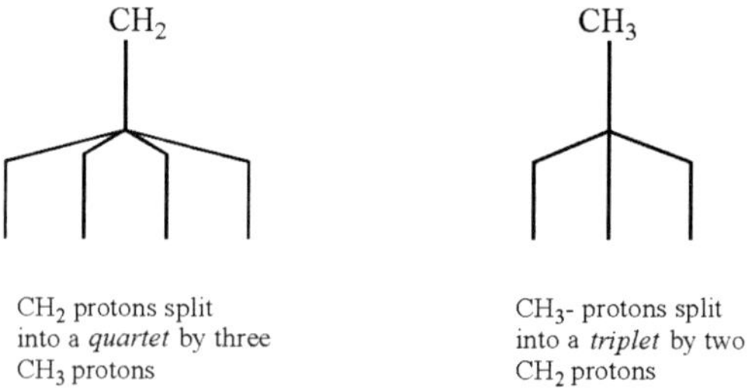

Figure 12.9. Spin-spin splitting in CH_3CH_2Cl.

Let us consider some spin-spin splitting patterns in aromatic compounds. All the protons in benzene are equivalent. So are the aryl protons in *para* disubstituted compounds like *p*-dimethyl benzene and *p*-dinitrobenzene. But in *p*-nitrotoluene, the two aryl protons near the methyl group are not equivalent to the two protons near the nitro group. The latter protons give signals downfield. In a monosubstituted aromatic compound, the protons of the ring are different, as given below.

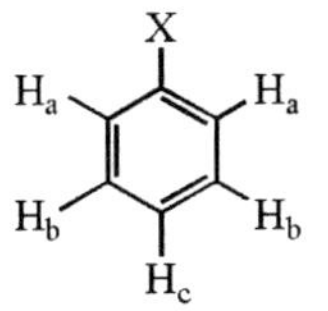

A monosubstituted compound

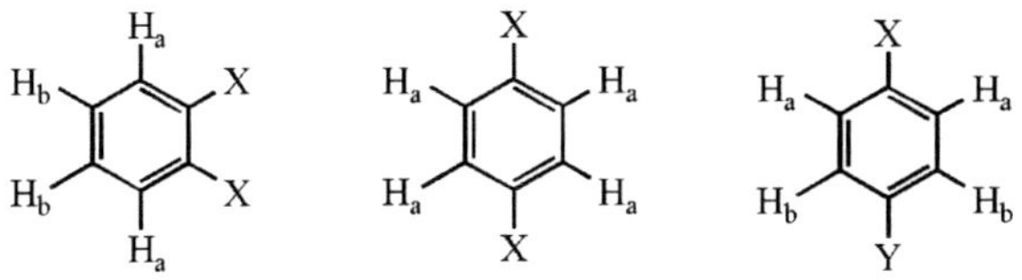

Disubstituted benzenes

1. X and Y are different substituents
2. H_a, H_b and H_c are non-equivalent protons

Coupling constant

The distance between the peaks in a multiplet is called the **coupling constant** (J) and is measured in hertz. Coupling constants vary from 0 to 18 Hz. The coupling constants of both sets of protons which are spin-split by one another are the same. For example, in the NMR spectrum of 2-chloropropanoic acid, $CH_3CHClCOOH$, there are two sets of signals; a doublet for the CH_3 protons split by the CH proton, and a quartet for the CH proton split by the three CH_3 protons. The CH_3 protons are coupled with CH protons, with a coupling constant of 8 Hz, and the CH protons are coupled with the CH_3 protons with the same coupling constant of 8 Hz.

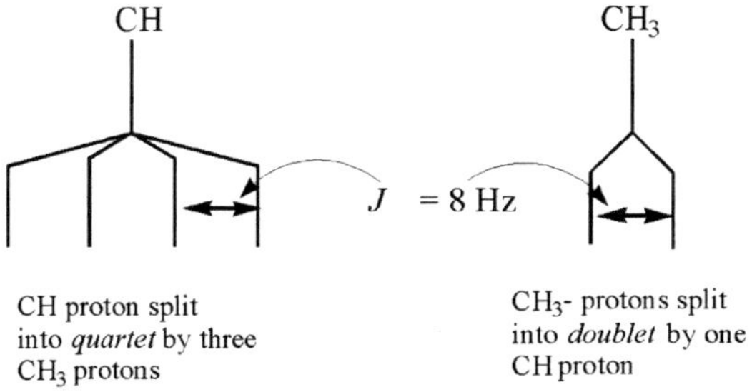

Figure 12.10. Spin-spin splitting in $CH_3CHClCOOH$.

Integration of NMR signals

The peaks are not of the same size in an NMR spectrum. The area under each peak is proportional to the number of protons. Thus, integrating the area under each peak and noting the height of the integrated line gives the number of protons.

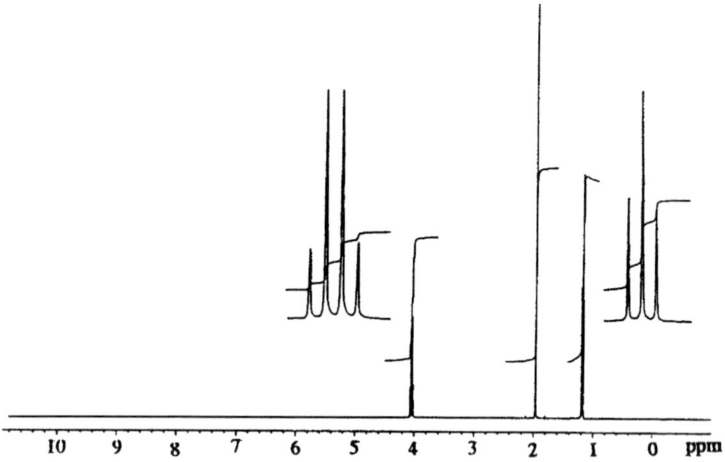

Figure 12.11. The ^1H NMR spectrum of ethyl acetate.

Let us examine the ^1H NMR spectra of two compounds. Figure 12.11 is the NMR spectrum of ethyl acetate ($CH_3COOCH_2CH_3$) and Figure 12.12, that of toluene ($C_6H_5CH_3$). The spectrum of ethyl acetate has three sets of peaks; a triplet at d 1.18 (integrates for three protons, $J = 7.2$ Hz, CH_3 protons of CH_2CH_3), a singlet at δ 1.97 (integrates for three protons, CH_3 protons of CH_3COO) and a quartet at δ 4.05 (integrates for two protons, $J = 7.2$ Hz, CH_2 protons of CH_2CH_3).

▶ *Key point* – The insets in Figures 12.11 and 12.12 show enlarged multiplets.

In the ^1H NMR spectrum of toluene, the three-proton singlet at δ 2.48 belongs to CH_3 group and the five-proton multiplet between δ 7.26 and δ 7.40, to C_6H_5 group. Since the protons of the C_6H_5 group are non-equivalent, they couple with each other to give the multiplet as can be seen in the diagram (H_a protons couple with H_b and H_c, H_b protons with H_a and H_c, and H_c proton with H_a and H_b).

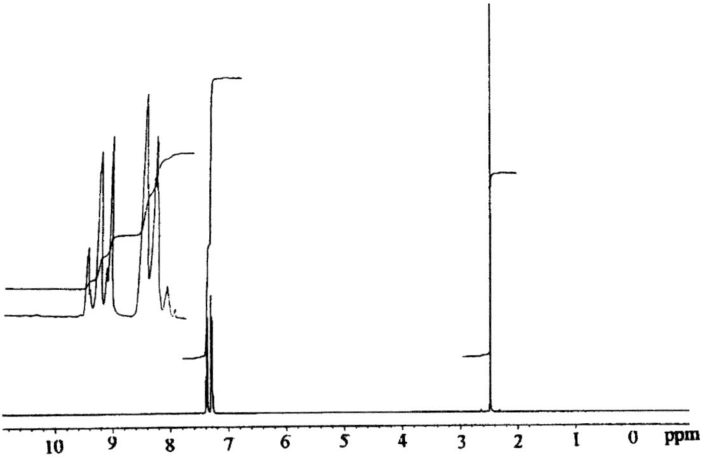

Figure 12.12. The ^{1}H NMR spectrum of toluene.

Summary

The NMR spectrum of a compound gives very important information about its structure.

1 From the position of the signals i.e. from the chemical shifts of the peaks, the type of protons in the molecule can be worked out (Table 12.3).

2 From the number of signals, the number of sets of equivalent protons can be deduced.

3 The integrals of the signals correspond to the number of protons in each signal.

4 The splitting pattern of each set of signals gives the number of nearby protons.

The ^{13}C NMR spectrum of a compound gives the number and type of carbon atoms in the molecule. ^{1}H and ^{13}C NMR spectra of compounds therefore provide very useful information about the structure of organic molecules.

Tutorial: helping you learn

Progress questions

1 Which of the following compounds give a λ_{max} in the 200-400 nm region of the UV spectrum:

(a) 1,3,5-Hexatriene (b) Phenylacetone

(c) 1,4-Hexadiene (d) 2,3-Dimethylbutane

2 Four compounds A, B, C and D give rise to characteristic peaks at wave numbers 1705, 1735, 2100 and 3300 cm^{-1} respectively. Assign functional groups for A, B, C and D from the list below:

(a) $-C \equiv C-$ (b) $>C=O$

(c) $-\overset{O}{\overset{\|}{C}}-OR$ (d) $-NH_2$

3 Write the formulae of the compounds listed in (a) to (e). How would you distinguish between these isomeric compounds by IR spectroscopy:

(a) Diethylether and 2-butanol.

(b) Butanoic acid and ethyl ethanoate.

(c) Butanone and 1-hydroxy-3-butene.

(d) 4-Methylbenzoic acid and methylbenzoate.

(e) 1,3-Butadiene and 1-butyne.

4 Predict the spin-spin splitting pattern expected, if any, in the ^1H NMR spectra of the following compounds:

(a) $CH_3\overset{O}{\overset{\|}{C}}-O\,CH_2\,CH_3$ (b) $CH_3OCH_2CH_2OCH_3$

(c) 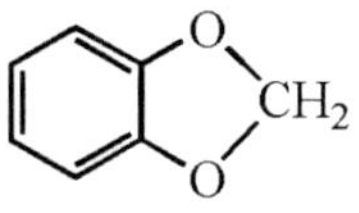

(d) 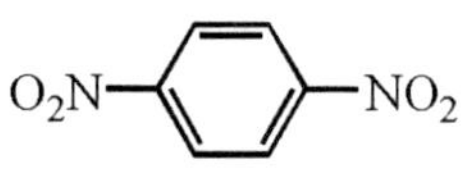

5 An aromatic compound of molecular mass 108 gives the following signals:

δ (ppm)	*Relative intensity*	*Multiplicity*
2.2	3	singlet
5.2	1	broad singlet
6.4	2	doublet
6.9	2	doublet

Propose a structure for the compound.

Glossary

achiral A molecule that is not chiral.

acyl halide A carboxylic acid derivative in which the -OH of the -COOH group is substituted with a halogen.

addition reaction A reaction in which a substance adds to another substance to form a compound.

alcohol A compound containing an -OH group attached to an sp^3 hybridised carbon atom.

aldehyde A compound containing a -CHO group.

aldol reaction An addition reaction in which two molecules of an aldehyde or a ketone combine to form a molecule of β-hydroxyaldehyde or β-hydroxyketone.

alkane A saturated, open-chained hydrocarbon.

alkene An unsaturated hydrocarbon containing a double bond in the molecule.

alkyne An unsaturated hydrocarbon containing a triple bond in the molecule.

amine An organic base formed by substituting one or more hydrogen atoms of ammonia by alkyl or aryl groups.

aprotic solvent A solvent like propanone or dichloromethane which has no hydrogen attached to its most electronegative atom.

aromatic compounds Compounds which contain one or more benzene rings.

atomic number The number of protons or the number of electrons in an atom.

atomic orbital The region of space where an electron is found.

Aufbau or building up principle In an atom, electrons occupy the innermost energy level, filling it before going to the next higher energy level.

bond dissociation energy The energy required to break the bonds in one mole of gaseous diatomic molecules.

bond length The distance between two atomic nuclei which share the bond.

Bronsted-Lowry acid A proton donor.

Bronsted-Lowry base A proton acceptor.

carbanion An ion which has a negative charge on a carbon atom.

carbocation An ion which has a positive charge on a carbon atom.

carboxylic acid A compound containing a -COOH group.

chemical shift The position of a signal in an NMR spectrum, expressed in parts per million (ppm) of the total applied magnetic field.

chiral carbon (asymmetric carbon) A carbon atom in a molecule which is bonded to four different atoms or groups of atoms.

chromophore The structural unit responsible for the UV absorption peak in a compound.

***cis-trans* isomers** Isomers which have different arrangement of their atoms in space due to the restricted rotation of a double bond or a ring.

conformation The different arrangements of atoms formed by the rotation of carbon atoms around a single bond.

constitutional isomers Isomers that have different arrangements of atoms in their molecules.

coupling constant (*J*) The distance between the peaks in a multiplet, measured in hertz.

covalent bond A chemical bond between two atoms in a molecule due to the attraction between the atomic nuclei and a pair of electrons shared by the two atoms.

cycloalkane A saturated hydrocarbon which has a carbon-carbon ring.

dehydrohalogenation A reaction in which a hydrogen and a halogen are eliminated from adjacent carbon atoms.

diastereomers Stereoisomers that are not mirror images of one another.

diol An alcohol which has two hydroxyl groups in a molecule.

dipole-dipole attraction Attraction between the δ+ve end of a dipole and the δ-ve end of another dipole.

eclipsed conformation A conformation where the carbon-hydrogen bonds are as close as possible.

electronegativity A measure of the power of an atom to attract the shared electron pair in a covalent bond.

electronic configuration An arrangement of available electrons of an atom in named atomic orbitals.

electrophile A molecule or a positive ion which can accept a pair of electrons from a donor and form a bond.

elimination A reaction in which two atoms or groups of atoms are removed from adjacent carbon atoms.

enantiomers Stereoisomers that are mirror images of one another.

enol An alcohol which has a hydroxyl group attached to a carbon of a carbon-carbon double bond.

free radical An atom (or a group of atoms which has an atom) with a single electron.

frequency The number of wave cycles that passes through a fixed point in unit time (usually per second). The unit is cycles per second (cps) or hertz (Hz).

Grignard reagent An alkyl or aryl magnesium halide of the type RMgX or ArMgX.

halogenoalkane A compound formed when one or more hydrogen atoms of an alkane are substituted by halogen atoms.

heat of combustion ($\Delta H^\circ C$) The heat evolved when one mole of a compound is burnt completely in air or oxygen.

homologous series A series of compounds with similar chemical properties and gradually varying physical properties, all the members of which can be represented by a general formula.

Hund's rule of maximum multiplicity When there is more than one orbital of equal energy value (for example, the three *p* orbitals in an energy level), electrons occupy them singly, before pairing takes place.

hybridisation The formation of a new set of atomic orbitals by the combination of two or more atomic orbitals.

hydrocarbons Compounds containing carbon and hydrogen only.

hydrogenation The addition of hydrogen to an alkene or an unsaturated compound.

hydrogen bonding Intermolecular forces of attraction between δ+ve hydrogen atoms bonded to strongly electronegative atoms (O, N, F or Cl) and such δ-ve electronegative atoms.

inductive effect The polarising effects in single covalent bonds is known as the inductive effect.

infrared spectroscopy A technique whereby the amount of infrared radiation a substance absorbs or transmits is measured and plotted on a graph to give an **infrared spectrum.**

isomers Different compounds with the same molecular formula.

ketone A compound containing a -CO group bonded to two carbon atoms.

Lewis acid A molecule or ion that can accept a pair of electrons to form a covalent bond.

Lewis base A molecule or ion that can donate a pair of electrons to form a covalent bond.

Markovnikov's rule It states that when a hydrogen halide or water reacts with an alkene, the hydrogen adds to the carbon of the double bond which is bonded to more hydrogen atoms.

meso compound A molecule which has two or more stereocentres, but is achiral because there is a plane of symmetry.

meta Refers to 1,3 positions of substituents on a benzene ring.

molecular orbital The region of space in a molecule where electrons are found.

monohydric alcohol An alcohol which has one hydroxyl group in a molecule.

nucleophile A molecule or an anion which has a lone pair of electrons that can form a bond with another atom or ion.

optical activity The property of enantiomers to rotate the plane of polarised light.
ortho Refers to 1,2 positions of substituents on a benzene ring.
para Refers to 1,4 positions of substituents on a benzene ring.
Pauli exclusion principle An orbital is occupied by a maximum of two electrons and these two electrons spin in opposite directions.
phenol A compound containing an -OH group attached to a benzene ring.
polarimeter A device used to measure the angle of rotation of plane-polarised light by an optically active compound.
precursor A compound that is converted into a target molecule.
primary carbon (1°) A carbon bonded to one alkyl group.
protic solvent A solvent like water or ethanol which has a hydrogen atom bonded to a very electronegative atom.
retrosynthesis or **retrosynthetic analysis** The process of reasoning backwards from the target molecule to the starting material(s).
saponification The alkaline hydrolysis of esters to give rise to alcohols and salts of carboxylic acids.
secondary carbon (2°) A carbon bonded to two alkyl groups.
sp^3 **hybridisation** The hybridisation of one *s* orbital and three *p* orbitals to form a set of four orbitals called sp^3 hybrid orbitals.
sp^2 **hybridisation** The hybridisation of one *s* orbital and two *p* orbitals to form a set of three sp^2 hybrid orbitals.
sp **hybridisation** The hybridisation of one *s* orbital and one *p* orbital to form a set of two *sp* hybrid orbitals.
staggered conformation A conformation where the carbon-hydrogen bonds are as far away from each other as possible.
stereoisomers Isomers with the same arrangement of atoms in their molecules, but differ in the arrangement of their atoms in space.
substitution reaction A reaction in which one atom or group of atoms is substituted by another atom or group of atoms.
tautomerism A type of isomerism exhibited by keto-enol tautomers.
tautomers Rapidly interconvertible constitutional isomers which differ in the position of a hydrogen and a double bond (Section 8.7).
tertiary carbon (3°) A carbon bonded to three alkyl groups.
unsaturated compound A compound which has two or three covalent bonds between two carbon atoms in a molecule.
van der Waals forces The forces of attraction between instantaneous dipoles (Section 3.2).
wavelength (λ) The length of one complete wave, from trough to trough or from crest to crest.
wavenumber ($\bar{\nu}$) The reciprocal of wavelength which is the number of waves per centimetre.

Answers to Practice Questions

Chapter 1

1. *Solution* 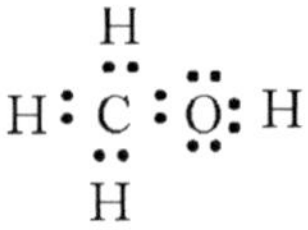

2. *Solution* 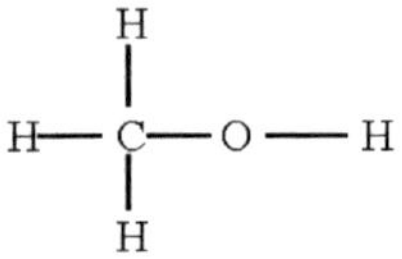

3. *Solution*

The carbon atom of the methyl (CH_3) group is sp^3 hybridised and the carbon atom of the nitrile (CN) group is sp hybridised. The H-C-H and the H-C-C bond angles are approximately 109.5° and the C-C-N bond angle is 180°.

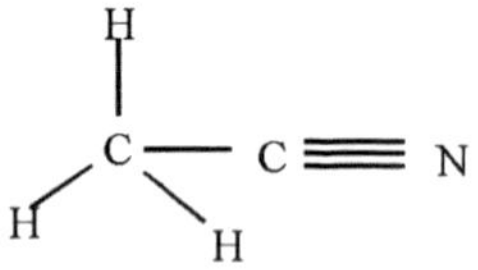

Chapter 2

1. *Solution*
   (a) 3-Chloro-2-ethylbutanoic acid
   (b) 3-Ethyl-4-methylhex-1-ene
   (c) Cyclohexyldimethylamine
   (d) 3,4-Dimethylpentan-2-ol

2. *Solution*

(a) 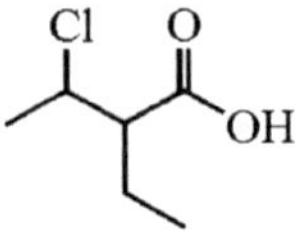

(b) 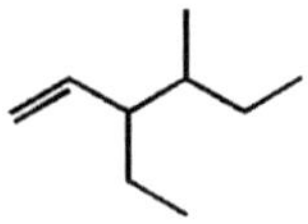

(c) 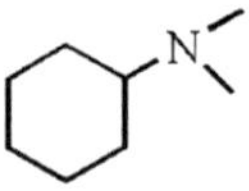

(d) 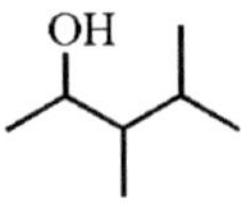

Chapter 5

1. $CH_3CH_2CH{=}CHCHOHCH_3$

2-Hydroxy-3-hexene

2.

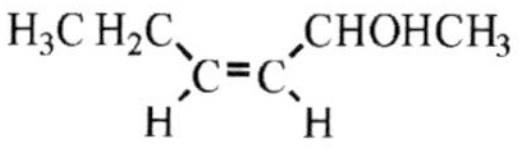

Z- isomer

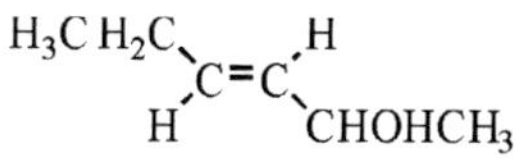

E- isomer

3.

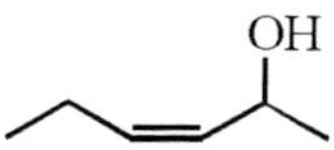

Z- isomer

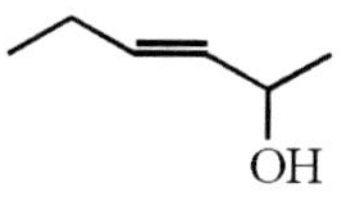

E- isomer

4.

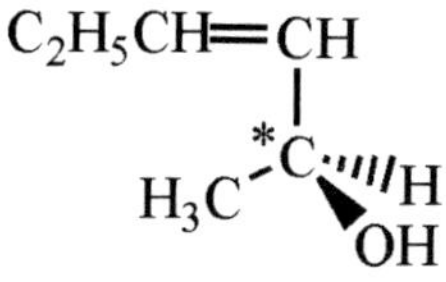

S- configuration

Chapter 11

(a) *Retrosynthetic analysis*

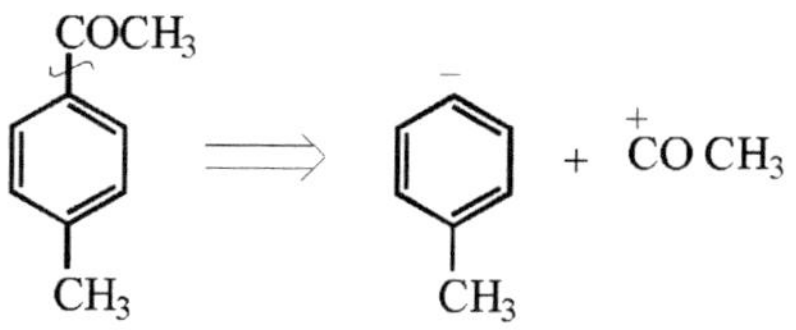

Synthesis

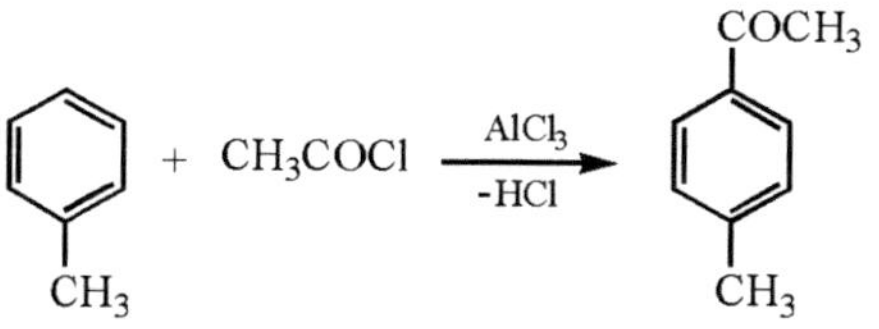

Toluene

4-Methylacetophenone

Toluene is easily available and Friedel-Crafts acylation with acetyl chloride gives rise to the target molecule.

(b) *Retrosynthetic analysis*

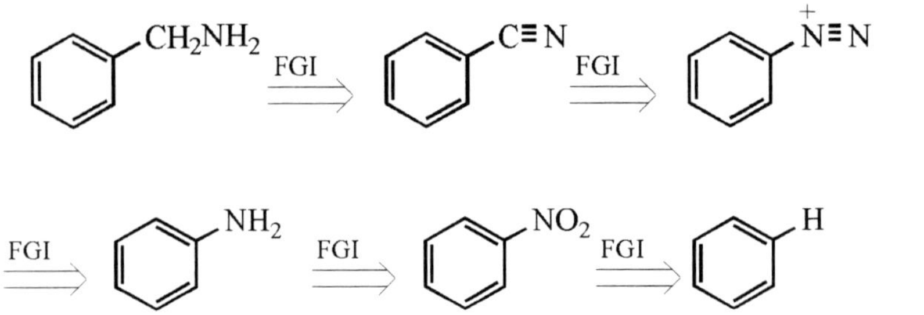

Synthesis

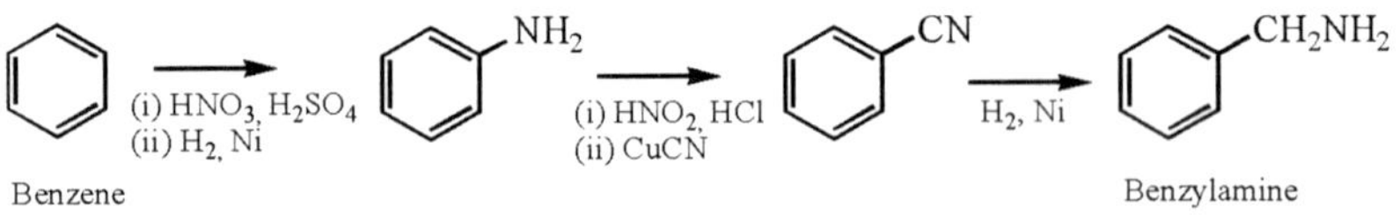

Web Sites for Chemistry Students

One minute summary – The internet, or world wide web, is an amazingly useful resource, giving the student nearly free and almost immediate information on any topic. Ignore this vast and valuable store of materials at your peril! The following list of web sites may be helpful for you. Please note that neither the author nor the publisher is responsible for content or opinions expressed on the sites listed, which are simply intended to offer starting points for students. Also, please remember that the internet is a fast-evolving environment, and links may come and go. If you have some favourite sites you would like to see mentioned in future editions of this book, please write to Dr Aleyamma Ninan c/o Studymates (address on back cover), or email her at the address shown below. You will find a free selection of useful and readymade student links for chemistry and other subjects at the Studymates web site. Happy surfing!

Studymates web site: http://www.studymates.co.uk
Aleyamma Ninan email: aninan@studymates.co.uk

Organic chemistry

Basic Reactions in Organic Chemistry
http://ourworld.compuserve.com/homepages/rehrler/
This interactive site uses VRML 2.0 techniques.

Bristol University School of Chemistry
http://www.bris.ac.uk/Depts/Chemistry/Bristol.Chemistry.html

Chemistry Help Online for Students
http://www.tznet.com/dnest/
This is an interactive organic chemistry learning tool for students

Classic Organic Reactions
http://home.ici.net/-hfevans/reactions.html
The site contains diagrams of more than 300 named organic reactions with references.

Infochem Organic Chemistry Courses
http://www.infochem.co.uk/courses/organic.htm
Infochem offers software and courses on everything from amino acids to common definitions and terms in organic chemistry.

Links for Chemists (University of Liverpool)
http://www.liv.ac.uk/Chemistry/Links/links.html
This excellent site contains over 7500 pointers to chemistry resources on the world wide web, including links to all UK university chemistry departments. Well worth a look.

Organic Chemistry
http://schmidel.com/bionet/o-chem.htm
The site offers a guide to the best biology and chemistry educational resources on the web: tutorials and references, molecular models and databases, organic molecules, problem sets and quizzes, and lots more.

Organic Chemistry Online
http://homework.chem.uic.edu
An interactive tutorial in sophomore-level organic chemistry featuring independent modules on key topics such as spectroscopy, stereochemistry, and reaction mechanisms.

Organic Chemistry Resources Worldwide
http://chemistry.gsu.edu/post.docs/koen/worgche.html
This is an intuitive web resource guide for synthetic organic chemists.

Royal Society of Chemistry
http://www.chemsoc.org/indexliv.htm
The leading UK professional and scientific body.

Sheffield's Chemdex
http://www.shef.ac.uk/-chem/chemdex/
Since 1993, Chemdex has maintained an international directory of chemistry on the internet. It contains more than 4,000 links, and is run from the University of Sheffield. It deals with organic chemistry at: http://www.chemdex.org/chemdex/organic.html

Undergraduate Organic Chemistry
http://interchem.chem.uab.edu/barbaro/
This University of Alabama site contains supplemental material on topics covered in undergraduate organic chemistry. It contains a guide to class handouts, tutorial sites, and more.

Search engines

Internet search engines have become big web sites in their own right. As well as search boxes, you will find handy directories of information, news, email and other services. There are thousands of search engines freely available, but the biggest and best known are probably AltaVista, Infoseek, Lycos and Yahoo!. Try out several and see which one(s) you like the best.

AltaVista
http://dir.altavista.com/Science/Chemistry/Organic.shtml
Alta Vista is one of the most popular search sites among web users everywhere. This is its portal to organic chemistry on the web.

Google
http://www.google.com
This simple-looking but powerful search engine is proving popular with regular users. A basic keyword search for 'organic chemistry' yielded more than 46,000 results.

HotBot
http://www.hotbot.com/
This is an impressive and well-classified search engine and directory.

Infoseek (Go Network)
http://infoseek.go.com/
Infoseek is another of the top ten search engines on the internet.

Internet Public Library
http://www.ipl.org/ref/
The 'ask-a-question' service at the Internet Public Library is experimental. The librarians who work here are mostly volunteers with full-time librarian jobs.

Internet Sleuth
http://www.isleuth.com
Internet Sleuth is a metasearch tool offering more than 3,000 databases to choose from.

List of Search Engines
http://www.merrydew.demon.co.uk/search.htm
This enterprising British site offers a free list of search engines, over 250 on different specialist topics. Well worth a look.

Lycos
http://www.lycos.co.uk/
Lycos is another of the top ten worldwide web search engines. This is its UK home page.

Metacrawler
http://www.metacrawler.com/
MetaCrawler collates results, eliminates duplication, scores the results and provides the user with a list of relevant sites.

Metaplus
http://www.metaplus.com/uk.html
Metaplus is a metalist of the best internet directories – and also offers direct links to some key general sites. This is its UK page, containing hundreds of classifications to explore.

Search.com
http://search.cnet.com/
This service is run by CNET, one of the world's leading new-media companies.

Search IQ
http://www.searchiq.com/
This site provides reviews of the most popular search engines and directories, including Altavista, AskJeeves, Excite, Google, Hotbot, Inference, Infoseek, Lycos, and Yahoo!.

Starting Point
http://www.stpt.com/search.html
This is a powerful metasearcher that puts hundreds of quality search tools at your fingertips.

Webcrawler
http://webcrawler.com/
Webcrawler is a fast worker and returns an impressive list of links, locating key words that may be buried deep within a document's text.

WebFerret
http://www.ferretsoft.com
WebFerret is not a search engine as such, but a handy little search utility. It queries ten or more leading search engines simultaneously, discarding duplicate results. You key in your query offline, and when you connect it searches the web until it has collected the number of references you have

specified – up to 9,999 if you wish. The program is free and only takes a few minutes to download. Highly recommended.

Yahoo! UK
http://www.yahoo.co.uk
Yahoo! was the first big search site on the internet, and is still one of the best for free general searching. It is probably one of the search engines and directories you will use time after time, as do millions of people every day.

Help with student research

Bath Information and Data Services
http://www.bids.ac.uk/
BIDS is a leading UK provider of networked information services for higher education and research. You can search 70,000 full text articles from biomedical, engineering, social sciences and other journals.

Britannica
http://www.eb.com
The web site of the famous encyclopaedia.

BT Campus World
http://www.campus.bt.com/campusworld/
This site indexes hundreds of other sites by subject. Mainly aimed at students at school, rather than at higher education, it might still contain things of value.

Bulletin Board for Libraries
http://www.bubl.ac.uk
BUBL is an excellent UK starting point for subject searches. It includes an alphabetic index and search tools.

Clearinghouse
http://clearinghouse.net/
This is a 'reference site of reference sites', divided into a large number of categories and sub-categories.

Edinburgh Data and Information Service
http://edina.ed.ac.uk
EDINA provides access to BIOSIS, with a broad coverage of life sciences, and an index to over 3,500 academic journals.

JANET
http://www.ja.net
If you are connecting from within an academic institution, an excellent starting point is the Joint Academic Network.

Learning Network
http://www.netlearn.co.uk
Here you will find learning resources, plus courses with tutor support.

Mailbase
http://www.mailbase.ac.uk
Mailbase is the place to explore more than 2,000 UK-based academic mailing lists.

MicroSoft for Students
http://msn.co.uk/default.asp
The software giant Microsoft does its bit for students. Visit this UK page and click on the Students link for access to UK news and articles, bulletin boards, news, and sites of student interest.

National Extension College
http://www.nec.ac.uk/cmain.html
The NEC offers a wide range of courses at levels from GCSE and NVQ to degree level, in an impressive number of subjects.

NISS
http://www.niss.ac.uk/
NISS provides keyword searching of some key UK higher education databases.

The Open University
http://www.open.ac.uk/
The first and most important distance learning course provider in the UK.

Topica
http://www.topica.com/
Topica (formerly Liszt) is the leading internet guide to more than 90,000 mailing lists, plus links to thousands of newsgroups and chat networks. See also *Mailbase.*

Index